CLEAR CURRENT

UNRAVELING THE WATER CRISIS

GIL D. BLUTRICH

FriesenPress

One Printers Way
Altona, MB R0G 0B0
Canada

www.friesenpress.com

Copyright © 2024 by Gil D. Blutrich
First Edition — 2024

ISBN
978-1-03-831138-2 (Hardcover)
978-1-03-831137-5 (Paperback)
978-1-03-831139-9 (eBook)

1. NATURE, ENVIRONMENTAL CONSERVATION & PROTECTION

Distributed to the trade by The Ingram Book Company

CLEAR
CURRENT
UNRAVELING THE WATER CRISIS

ABOUT GIL BLUTRICH

Gil Blutrich was born in 1965 in Tel Aviv, Israel. From a young age, he showed a knack for entrepreneurship. At the age of sixteen, he launched his first business venture, making him the youngest person in Israel registered for income tax during that period. After completing his high school education, Blutrich served in the Israeli Air Force, a tenure that instilled in him discipline and leadership qualities. Post-military service, he pursued higher education in hotel management, laying the groundwork for his future career in real estate and hospitality.

In 1990, Blutrich founded Mishorim, a real estate company in Israel. His leadership steered Mishorim to significant growth, leading to its public listing on the Tel Aviv Stock Exchange in 2007. Eight years prior, in 1998, he moved to Canada and established Skyline Investments, focusing on the hospitality sector in North America. Skyline Investments saw its public listing in 2014.

Blutrich has been instrumental in various significant real estate projects and hospitality ventures. These include the development and refurbishment of landmark properties like the King Edward Hotel in Toronto and the Renaissance Hotel in Cleveland. His influence extends across diverse geographical locations, including Israel, the United States, and Canada.

Beyond his business achievements, Blutrich is known for his commitment to social causes. He founded the social business initiative "Doing Business," which focuses on teaching entrepreneurship to at-risk youth across over fifty communities in Israel. He has also served as a board member of the Canada-Israel Chamber of Commerce, fostering business and cultural ties.

Blutrich's business expertise has earned him several accolades, including the Entrepreneur of the Year in Ontario's hospitality sector in 2004. His companies have been listed among the top fifteen in Canada's Best Managed Companies program by Deloitte, a testament to his exceptional leadership and business acumen.

As the honorary consulate of El Salvador in Canada, Blutrich has played a role in promoting cultural exchange and strengthening bilateral relations between Canada and El Salvador. He was instrumental in the return of the historic steamship *SS Keewatin* to Canada. He also chaired the Israeli Forum in Canada for five years, enhancing the Israeli community's presence and interests in Canada.

Mr. Blutrich acts as business development advisor to his family holdings office. Among his family holdings are commercial and residential holdings and large-scale development sites in the US and Canada.

In his latest venture, Blutrich co-founded Clear Inc., focusing on transforming buildings and communities into healthier and more sustainable environments. This initiative reflects his commitment to environmental sustainability and innovative technology in water management.

Gil Blutrich's story from a young entrepreneur to a globally recognized business leader and philanthropist illustrates his relentless pursuit of innovation, excellence, and social responsibility. His diverse contributions to various sectors and his ongoing efforts in environmental sustainability continue to inspire many around the world.

SPECIAL ACKNOWLEDGEMENTS

Before diving into this narrative, I must express my heartfelt gratitude to those who have been instrumental in my journey. To my mother, Tova, whose strength and wisdom have always guided me; my sister, Adi, a constant source of support; and in loving memory of my late father, Uzi Blutrich, whose legacy continues to inspire me. To my partner, Natalia, who has been a pillar of support, sharing in the vision of making a positive impact.

To my older children, Ben and Ron, standing beside me, who contribute significantly to our mission of making the world a better place. A special thank you to Yariv Avriamovitz, CEO of Atlantium Technologies, for opening doors of knowledge and opportunity in my quest to understand and tackle the water crisis.

INTRODUCTION

For over three decades, my life as a real estate entrepreneur, hotelier, and developer has been about more than constructing and retrofitting buildings; it has been about building communities and enriching lives. My real estate journey across over fifty different cities in Israel, Canada, and the US has taught me the importance of creating spaces that do more than just exist—they thrive and enhance the day-to-day experiences of their inhabitants.

My path to addressing the global water crisis began with a personal event—my son Ron's struggle with a waterborne illness. This incident shifted my focus from just creating better living spaces to ensuring the fundamental health and safety of my family, tenants, and clients. It was a turning point that propelled me into the realm of environmental health, particularly water safety.

The global water crisis is a complex, multifaceted issue. It's not just about water scarcity; it's about access, quality, and sustainability. This crisis affects billions of people worldwide, undermining health, education, and economic development. As I delved deeper into this issue, the magnitude of the challenge became increasingly evident, transcending geographical and socio-economic boundaries.

Guided by the Jewish principle of Tikkun Olam—"repairing the world"—I embarked on this mission to contribute positively to our planet. This journey is an embodiment of Tikkun Olam, aligning my professional skills and personal values toward a cause that extends beyond individual interests.

My explorations led me to discover the transformative potential of UV water purification technology—a beacon of hope in addressing waterborne diseases. However, this book aims to go beyond UV technology, exploring a range of innovative solutions that are reshaping our approach to water challenges, from advanced filtration systems to intelligent water management.

Clear Current is a call to action—an invitation to join me in understanding and addressing the water crisis. It is a journey of discovery, responsibility, and hope, urging each of us to play our part in securing clean and safe water for all.

As we embark on this expedition through these

pages, I am reminded of the collective responsibility we share in securing a sustainable water future. This book is my contribution to a global conversation, a narrative that intertwines personal reflections with a broader vision for sustainable times ahead. Together, let's create a world where clean and safe water is not a luxury, but a fundamental right for all.

Gil D. Blutrich
Jan. 2024

TABLE OF CONTENTS

CHAPTER 1: UNDERSTANDING THE WATER CRISIS

My journey into understanding the water crisis began through personal encounters and travels. Witnessing the struggles of communities in drought-stricken and flood-affected regions opened my eyes to the stark reality of water scarcity. These experiences transformed my understanding and approach toward water, stirring a sense of responsibility and a desire to advocate for change. The intersection of water scarcity and human rights became a pivotal aspect of my reflections, and I've been inspired by innovations and solutions emerging around the world. From community-led initiatives to technological advancements like ultraviolet water purification, there are tangible hopes for addressing this crisis.

I've learned that tackling the water crisis requires collaboration across sectors and borders. Sharing knowledge, resources, and best practices is crucial for addressing water scarcity. This crisis isn't just about water; it's about human dignity, equity, and sustainability.

Reflecting on these experiences, I've realized the profound impact access to clean water has on communities. It's not just about constructing buildings or communities; it's about ensuring fundamental health and safety. This realization shifted my perspective and approach to development, making me more mindful of how my projects can contribute to water sustainability.

Seeing the water crisis through the lens of human rights has been key. It's not just an environmental or economic issue; it's a matter of human dignity. It is imperative to make certain that water access is intrinsically linked to upholding human rights and fostering equitable development.

Despite the crisis's severity, I remain inspired by the resilience of affected communities and the innovative solutions emerging globally. My journey into the water crisis has been both a personal and a collective one, reinforcing my commitment to sustainable development and responsible stewardship of our planet's resources. As we move forward, it's essential that we all work together to ensure a water-secure future for all.

I believe that the water crisis is a defining challenge of our time, transcending geographical boundaries and impacting billions worldwide. According to a 2019 joint report by UNICEF and the World Health Organization (WHO), 2.2 billion people lack safely managed drinking water services. This means that one in three people globally do not have access to safe drinking water, underscoring a significant failure in basic service provision and highlighting deeper issues of inequality and human rights neglect. By 2025, an estimated 1.8 billion people will face water scarcity, with two-thirds of the world's population potentially in water-stressed conditions. This scenario, hastened by climate change and population growth, is a rapidly approaching reality.

According to the 2019 report by UNESCO, In Sub-Saharan Africa, only 24 percent of the population has access to safe drinking water, compared to over 90 percent in developed regions like Europe and North America, reflecting global socio-economic disparities. Water scarcity appears as physical scarcity in regions like sub-Saharan Africa, worsened by climate change and environmental factors, and as economic scarcity in South Asia, where inadequate infrastructure hinders water access despite natural availability.

The effects of water scarcity are profound, particularly on women and children, who bear the brunt of water collection responsibilities, often at the cost of education or economic activities. Population growth, urbanization, and increasing agricultural demands intensify competition for water resources, escalating conflicts and stress on water systems.

Water scarcity impacts regions worldwide, each facing unique challenges:

- Sub-Saharan Africa: struggles with severe droughts and water shortages, exacerbated by political instability and inadequate infrastructure. (1)
- Middle East and North Africa: face water scarcity intensified by arid climates and political conflicts over shared water resources. (2)
- South Asia: grapples with both physical and economic water scarcity, worsened by rapid urbanization and industrialization. (3)
- The Americas: encounters diverse water challenges, from droughts in California to inequitable access in Latin America. (4)
- Australia: deals with prolonged droughts and climate change effects, leading to significant declines in water availability. (5)

Water scarcity has far-reaching socio-economic implications

The water scarcity crisis brings with it an often-overlooked yet severe consequence: the proliferation of waterborne diseases. This issue straddles the divide between the Global South and developed nations, including North America. The scarcity and consequent reliance on unsafe water sources set the stage for a range of diseases, from cholera to dysentery and typhoid, posing significant public health risks.

In regions like sub-Saharan Africa, the intermittent and unreliable access to clean water forces many to depend on contaminated sources. This reliance on unsafe water leads to a high prevalence of cholera, a disease marked by intense diarrhea and dehydration. In areas with limited healthcare resources, cholera cases can quickly escalate into a public health emergency. According to WHO, the global burden of cholera is estimated at 1.3 to 4.0 million cases each year, resulting in 21,000 to 143,000 deaths. These figures reflect a stark reality of the health implications of water scarcity.

Beyond cholera, other diseases like dysentery and typhoid fever are equally concerning. Dysentery, caused by bacterial or parasitic agents, manifests in severe gastrointestinal distress and can be fatal if untreated. The Global Burden of Disease Study reported around 125,000 deaths globally due to bacterial dysentery in 2019. Typhoid fever, another disease linked to poor water and sanitation conditions, affects millions. The CDC highlights that there are approximately 21 million cases of typhoid fever annually, leading to 161,000 deaths. Such diseases disproportionately impact the most vulnerable population segments, including children, the elderly, and those with compromised immune systems.

In North America, waterborne diseases present a different set of challenges than other regions of the world, but they are equally alarming. Cases arising from pathogens such as Legionella, Giardia, and Norovirus, often linked to contaminated drinking water, underline the vulnerability of even advanced water systems. The Flint water crisis in Flint, Michigan, US, is a notable example, where a series of systemic failures led to widespread lead exposure and outbreaks of Legionnaires' disease. This crisis highlighted how infrastructure lapses and regulatory oversights could lead to public health disasters.

Climate change further compounds the risk of waterborne diseases in North America. Increasingly frequent and severe weather events, such as hurricanes and floods, exacerbate the potential for water contamination. The aftermath of Hurricane Katrina in New Orleans exemplifies this risk, where floodwaters led to widespread contamination of the water supply, resulting in a spike in gastrointestinal illnesses. These incidents underscore the intricate link between environmental factors, infrastructure integrity, and public health.

The issue of waterborne diseases in the context of water scarcity is a clear indicator of the interconnectedness of global health, environmental conditions, and socio-economic factors. While the manifestation of these diseases varies between regions like sub-Saharan Africa and North America, the underlying challenge remains consistent: ensuring access to safe, clean water. As the world grapples with the dual challenges of water scarcity and climate change, the need for concerted efforts to safeguard public health against the threat of waterborne diseases becomes ever more pressing.

CHAPTER 2: CAUSES AND CONSEQUENCES

Four primary causes of the water crisis include climate change, overuse and mismanagement of water resources, and pollution.

Climate Change: A Journey Through the World's Shifting Waters

In my travels and research, I've witnessed firsthand the profound and deeply personal impact climate change is having on water availability around our planet. It's a story of contrast and extremes—a narrative deeply etched in the lives of countless individuals and communities I've encountered.

Journeying through various landscapes, I've seen how altered precipitation patterns disrupt lives. In some parts of the world, what used to be lush, predictable rainy seasons have turned into sporadic, inadequate drizzles, leaving fields parched and wells dry. In contrast, other regions, which once enjoyed moderate rainfall, now face torrential downpours that lead to devastating floods. Each drop of rain, or its absence, tells a tale of changing times and shifting fates.

In places like the sprawling savannas of Africa and the vast outbacks of Australia, the increased frequency of droughts is not just a climatic event but a profound life-altering reality. I've sat with farmers staring helplessly at the sun-cracked earth, once fertile, now barren. Their stories are not just about lost crops but about traditions, heritage, and a way of life withering away under the relentless sun.

Floods – The Overflowing Anguish: The flip side of this climatic coin is the sudden, often furious, onset of floods. Cities, where life once bustled, now lie submerged and silent. The floodwaters are indiscriminate, sweeping away both the memories of the past and the dreams for the future. Witnessing the aftermath of these floods, I've heard heartwrenching stories of loss and resilience, of lives uprooted and the struggle to rebuild in the wake of nature's fury.

Glaciers – A Melancholic Farewell: Perhaps nothing has been more poignant in my journey than witnessing the retreat of majestic glaciers. These frozen titans, once thought eternal, now weep rivers of melted ice. Communities that have thrived on their life-giving waters now face an uncertain future. Each drop from the melting glaciers is a ticking clock, a reminder of the transient nature of our environment.

Ecosystems in Peril: Beyond the human toll, the ecological impact is stark. The changing water scenarios are altering habitats at a pace too rapid for many species to adapt. From the vibrant coral reefs to the dense rainforests, the diminished water resources are reshaping ecosystems, often with irreversible consequences.

In documenting these stories, my aim is not just to chart the physical changes in our environment but to capture the human emotion and resilience behind these shifts. Climate change and its impact on water is not just a tale of statistics and forecasts; it's a human story, woven into the very fabric of our existence. As I continue my own journey, the urgency for a united, global response to these challenges becomes ever clearer. The future of our planet, our water, and our collective existence depends on it.

Overuse and Mismanagement

In the tapestry of our global water crisis, two threads are consistently prominent yet often overlooked: overuse and mismanagement. These twin forces, manifesting across varied landscapes—from sprawling agricultural fields to bustling urban centres— weave a complex pattern of scarcity and waste.

Agriculture

As I explore the complexities of water usage, it becomes clear that agriculture is a dominant force, consuming about 70% of the planet's freshwater resources. The Food and Agriculture Organization of the United Nations notes the irony in this, highlighting the significant inefficiencies in agricultural water use.

Inefficient Irrigation: I've witnessed fields inundated with water, far more than crops require. Traditional irrigation methods like flood or furrow irrigation are widespread, yet startlingly, only about 35–50 percent of this water actually serves the intended purpose. The rest? It's lost—a silent witness to evaporation, runoff, or percolation into the unseen depths.

Thirsty Crops: The water footprint of crops like cotton, sugarcane, and rice is staggering. To put it in perspective, producing just one kilogram of cotton can consume about 10,000 litres of water. It's a thirsty affair, often taking place in regions already gasping for water.

Runoff and Pollution: As I explore the consequences of agricultural runoff, laden with fertilizers and pesticides, the impact on water bodies is glaring. This runoff doesn't just deplete water quality but also triggers ecological imbalances like algal blooms, disrupting aquatic life.

Industry

The industrial sector, a behemoth in its own right, also guzzles vast quantities of water.

High Consumption Industries: My journey into industrial zones reveals that sectors such as textiles, paper, and chemicals are among the heaviest consumers of water. For example, the production of a single ton of paper can require up to 60,000 litres of water.

Pollution from Wastewater: The aftermath of industrial processes is often a cocktail of polluted water. Without proper treatment, this wastewater is released into local water bodies, contributing to a mounting crisis of water pollution. The World Bank's estimate that 70 percent of industrial wastes in developing countries are discharged untreated into waters only exacerbates the problem.

Urban Development

The "leaky cities" urban landscapes, epicentres of human congregation, present a unique set of challenges in water management.

Household Wastage: In many households, the story of water is one of unnoticed waste. A single leaky faucet, dripping away day and night, can waste up to 5,500 litres of water annually.

Crumbling Infrastructure: My exploration of urban water systems reveals a startling fact: in North American cities, about 30 percent of water never reaches the taps it's intended for. It's lost, seeping out from aging and leaking pipes.

Stormwater Woes: The management of stormwater, especially in cities with vast concrete expanses, often falls short, leading to polluted waterways and overloaded sewage systems.

As I reflect on these sectors, it becomes increasingly clear that the water scarcity crisis is as much about mismanagement and overuse as it is about physical scarcity. The challenge is immense, but understanding these dynamics is a crucial step toward navigating the path to a more sustainable and water-secure future.

Pollution's Grip on Our Water

In the global dialogue about water scarcity, pollution emerges as a silent yet formidable foe. It manifests through varied sources, each contributing to the degradation of our precious freshwater resources. The consequences of this pollution are far-reaching, affecting everything from public health to food security and economic stability.

Industrial Discharge: industrial activities are significant contributors to water pollution, and they are leaving a toxic legacy.

Annually, industries discharge millions of tons of pollutants into water bodies. For instance, it's estimated that textile dyeing and treatment contribute up to 20 percent of global industrial water pollution.

Heavy metals such as mercury, lead, and cadmium, commonly found in industrial effluents, pose severe health risks. These metals can accumulate in the ecosystem, entering the human food chain through fish and other aquatic life.

The chemical and pharmaceutical industries are notorious for releasing compounds that are often not effectively removed by conventional water treatment processes.

Agricultural Runoff: though vital for food production, this plays a significant role in water contamination. Nonpoint source pollution from agricultural fields, including excess fertilizers and pesticides, is a major concern. Nitrate pollution from fertilizers is particularly problematic, leading to groundwater contamination and eutrophication in surface waters.

The United States Geological Survey (USGS) estimates that agricultural runoff is the leading source of pollutants in rivers and streams, the second most significant source in wetlands, and the third main source in lakes.

Urban Pollution's Contribution: In urban areas, improper waste disposal exacerbates water quality issues. Untreated or inadequately treated sewage is a primary urban pollutant. The World Health Organization (WHO) estimates that around 1.8 billion people globally use a drinking water source contaminated with fecal matter.

In many developing countries, a significant

proportion of urban sewage—over 80 percent in some cases—is discharged untreated into rivers, lakes, and coastal areas. The impact of pollution on water scarcity is multidimensional:

Public Health Risks: Contaminated water is a leading cause of diseases such as cholera, dysentery, and typhoid. According to the WHO, about 829,000 people are estimated to die each year from diarrhea as a result of unsafe drinking water, sanitation, and hand hygiene.

Impact on Food Security: Water pollution directly affects agriculture, jeopardizing food production. Crop yields diminish due to polluted irrigation water, and contaminated water leads to seafood that is unsafe for consumption.

Economic Impact: The economic cost of water pollution is substantial. The World Bank estimates that the annual economic loss due to poor sanitation and hygiene amounted to $260 billion globally in 2010.

Social and Political Tensions: water scarcity exacerbated by pollution can lead to social unrest and political tensions, especially in regions where access to clean water is a key survival issue.

In summary, pollution from industrial discharge, agricultural runoff, and improper waste disposal are significant contributors to global water scarcity. Its impact is far-reaching, affecting public health, food security, and economic stability, often fueling social and political tensions. Addressing this visible crisis requires a comprehensive understanding of its sources and a concerted effort to mitigate its effects.

Footnotes:

(1): Intergovernmental Panel on Climate Change. "Climate Change 2021: The Physical Science Basis."

(2): United Nations Office for Disaster Risk Reduction. "Global Assessment Report on Disaster Risk Reduction 2022."

(3): World Wildlife Fund. "Living Planet Report 2022: Building a nature-positive society."

(4): United Nations World Water Development Report 2020: Water and Climate Change. UNESCO, 2020.

CHAPTER 3: THE PUBLIC HEALTH CRISIS OF WATER CONTAMINATION

Introduction to Waterborne Diseases

Waterborne diseases represent a significant threat to global public health, impacting millions of lives each year. These diseases are primarily caused by pathogenic microorganisms present in contaminated water, leading to a wide range of illnesses. Diarrheal diseases, including cholera, dysentery, and typhoid fever, are among the most common and deadly. According to the World Health Organization (WHO), diarrheal diseases account for an estimated 1.5 million deaths annually, predominantly in children under five in developing countries. (1)

Contaminated water can also lead to diseases like hepatitis A, polio, and cryptosporidiosis. These illnesses vary in severity and symptoms, but they share a common source: the ingestion of water contaminated with fecal matter, pathogens, or harmful chemicals. The impact of these diseases is widespread, extending beyond immediate health effects to long-term consequences like malnutrition, stunted growth, and impaired cognitive development.

Impact on Children's Health

Children are particularly susceptible to waterborne diseases due to their developing immune systems and higher exposure risks, especially in areas with poor sanitation and hygiene practices. Young children, especially those in low-income countries, bear the brunt of the health burden from contaminated water. Diarrheal diseases, for instance, are a leading cause of death in children under five, causing more fatalities than malaria, measles, and AIDS combined. (2)

The impact of these diseases on children's health is profound. Frequent or severe bouts of diarrhea lead to malnutrition and dehydration, which can have lasting effects on a child's growth and development. Furthermore, chronic exposure to waterborne pathogens can lead to conditions like stunting and cognitive impairment, hampering educational achievement and future economic productivity.

Epidemic Outbreaks and Case Studies

Waterborne disease outbreaks can have devastating effects, often exacerbated in areas with limited access to healthcare and sanitation infrastructure. Case studies of significant outbreaks illustrate these impacts:

Cholera in Haiti (2010): Following the 2010 earthquake, Haiti experienced a massive cholera outbreak, the first in the country in over a century. Poor sanitation conditions and limited access to clean water facilitated the rapid spread of the disease. The outbreak resulted in over 800,000 cases and nearly 10,000 deaths, highlighting the critical need for safe water and sanitation in disaster response. (3)

Flint Lead Contamination (2014): In Flint, Michigan, US, a change in water supply led to lead contamination, exposing residents to toxic levels of lead. Children were particularly affected, with potential long-term health consequences including neurological damage and developmental delays. (4)

The Walkerton, Ontario E. coli outbreak in May 2000: was a major public health disaster caused by contamination of the town's water supply. Heavy rainfall washed manure from a nearby cattle farm into one of the town's wells, introducing E. coli O157 and Campylobacter jejuni bacteria into the drinking water system. As a result, seven people died and over 2,300 fell ill, making it the most severe case of waterborne disease in Canadian history. The outbreak was exacerbated by improper water treatment and inadequate testing procedures by local authorities. A subsequent public inquiry led to significant changes in water safety regulations and practices in Ontario

Cryptosporidiosis in Milwaukee (1993): One of the largest waterborne disease outbreaks in US history occurred in Milwaukee when over 400,000 people were infected with Cryptosporidium from the municipal water supply. The outbreak led to widespread illness and over fifty deaths, underscoring the importance of water treatment and pathogen detection. (5)

These case studies demonstrate the catastrophic potential of waterborne diseases and the crucial role of robust water treatment and monitoring systems in preventing epidemics.

Unreported Burdens and the "Iceberg" Effect

The full impact of waterborne diseases is often significantly underreported, a phenomenon sometimes referred to as the "Iceberg Effect." This term suggests that for every case of waterborne illness that is diagnosed and reported, there are many more that go unrecognized or unrecorded. This underreporting can be attributed to several factors, including limited healthcare access, lack of diagnostic facilities, and asymptomatic or mild cases that are not medically treated.

In many developing countries, the burden of waterborne diseases is particularly obscured due to inadequate surveillance and reporting systems. The consequence is a significant gap in understanding the true scale and impact of these diseases. This lack of visibility impedes effective public health interventions and resource allocation, allowing the cycle of contamination and disease to continue unchecked.

CDC's Role in US Water Safety

The Centers for Disease Control and Prevention (CDC) plays a crucial role in monitoring and ensuring water safety in the United States. The CDC works in conjunction with local and state health departments to track waterborne disease outbreaks, investigate their sources, and implement measures to prevent future occurrences. They provide guidance on water treatment, safe drinking water practices, and emergency response to contamination events.

One of the key initiatives of the CDC is the Waterborne Disease and Outbreak Surveillance System (WBDOSS), which collects data on waterborne disease outbreaks associated with recreational water, drinking water, and environmental and unde-termined exposures. This surveillance system helps identify trends, high-risk factors, and effective pre-vention strategies, shaping water safety policies and practices across the nation.

Global Efforts in Improving Water Quality

Improving water quality on a global scale involves coordinated efforts from international organiza-tions, governments, NGOs, and local communities. Initiatives such as the United Nations' Sustainable Development Goal 6 aim to ensure the availability and sustainable management of water and sanita-tion for all. This goal encompasses targets related to drinking water quality, wastewater treatment, and protecting and restoring water-related ecosystems.

Organizations like the WHO and UNICEF play a pivotal role in these global efforts. They work to implement water safety plans, improve sanitation facilities, and enhance water quality monitoring and reporting in developing countries. However, challenges such as lack of funding, political instability, and inadequate infrastructure often impede progress in many regions.

WASH Programs and Their Importance

Water, Sanitation, and Hygiene (WASH) programs are critical in the fight against waterborne diseases. These programs focus on providing access to safe water, improving sanitation facilities, and promoting good hygiene practices, particularly in low-income and vulnerable communities. WASH initiatives have shown to significantly reduce the incidence of waterborne diseases, improve nutritional outcomes, and enhance the overall health and well-being of communities.

Organizations like UNICEF and the World Bank have been at the forefront of implementing WASH programs globally, emphasizing their importance in schools, healthcare facilities, and emergency contexts. WASH programs are also vital in empowering women and girls, who are often disproportionately affected by the lack of water and sanitation facilities.

Policies and Actions for Safer Water

Developing and implementing effective policies and actions is essential for ensuring safer water and combating waterborne diseases. This involves investing in water treatment infrastructure, enforcing strict regulations on industrial and agricultural water pollution, and promoting efficient water use and conservation practices.

Public awareness campaigns and education are also crucial in preventing waterborne diseases. Informing communities about safe water practices, the risks of contaminated water, and the importance of hygiene can significantly reduce disease transmission. Moreover, fostering community engagement in water management and decision-making processes ensures that solutions are sustainable and culturally appropriate.

The integration of technology, such as remote sensing for water quality monitoring and mobile applications for reporting and education, is also an emerging strategy in improving water safety. These technological advancements can enhance data collection, facilitate timely interventions, and support community-led monitoring efforts.

Footnotes:

(1): World Health Organization. "Diarrhoeal disease."

(2): UNICEF. "Water, Sanitation and Hygiene."

(3): Centers for Disease Control and Prevention. "2010–2019 Cholera in Haiti."

(4): Hanna-Attisha, Mona, et al. "Elevated Blood Lead Levels in Children Associated with the Flint Drinking Water Crisis: A Spatial Analysis of Risk and Public Health Response." American Journal of Public Health, vol. 106, no. 2, 2016, pp. 283–290.

(5): Mac Kenzie, W. R., et al. "A Massive Outbreak in Milwaukee of Cryptosporidium Infection Transmitted through the Public Water Supply." New England Journal of Medicine, vol. 331, no. 3, 1994, pp. 161–167.

CHAPTER 4: CURRENT SOLUTIONS AND THEIR LIMITATIONS

I believe it is essential to acknowledge the rich history and evolution of this vital field. From ancient civilizations utilizing sand filtration, boiling, and copper vessels, to the sophisticated technologies of our era, the journey of water purification is a testament to human ingenuity and the enduring quest for clean water.

In my exploration of this area, I've discovered that the methods of the past have not only survived but have evolved and integrated into the complex systems we see today. This progression reflects our growing understanding of waterborne contaminants, the health risks they pose, and our relentless pursuit of solutions.

Today, driven by technological advancements and

heightened environmental and health awareness, we've developed a wide array of techniques to purify water. Each of these methods, with their unique mechanisms and applications, represents humanity's multifaceted approach to tackling the diverse challenges in water purification. In my research, I've identified fifteen current procedures that stand as significant milestones in this journey. These methods are not merely solutions but also symbolize our adaptive strategies in the face of constantly evolving environmental and industrial contexts.

In this chapter, I will provide an in-depth exploration of these fifteen practices, examining their applications, advantages, limitations, and respective impacts on the market of water purification. My aim is to present a holistic view that recognizes both the technological achievements and the challenges that remain. This understanding is crucial for paving the way for future innovations and improvements in our ongoing quest for clean and safe water.

MECHANICAL FILTRATION

Applications

This method is extensively used in both residential and industrial settings.

In households, it is commonly found in water

filtration systems, serving as the first line of defense by removing visible particles and sediments.

Industrially, it's used in large-scale water treatment plants, manufacturing processes, and even in aquatic systems like fish tanks and ponds. (2) Mechanical filters are also vital in preliminary treatment steps for wastewater before it undergoes more intensive purification.

Advantages

Simplicity: one of the simplest forms of filtration, easily understandable and maintainable.

Cost-Effectiveness: Generally, it's more affordable than other advanced filtration systems. The maintenance and replacement of filter media are also relatively inexpensive.

Efficiency: highly effective for removing large particulate matter, thus protecting subsequent, more sensitive filtration systems from clogging or damage.

Versatility: can be used in a variety of settings, from small-scale home systems to large municipal water treatment facilities.

Limitations

Inability to Remove Microscopic Contaminants:
mechanical filters cannot remove chemical pollutants, dissolved solids, or microorganisms such as bacteria and viruses.

Maintenance Requirements: over time, the filter media can get clogged with particles, necessitating regular cleaning or replacement.

Water Quality Dependency: The effectiveness can vary depending on the quality of the incoming water. Highly turbid or contaminated water can reduce the efficiency and lifespan of the filter.

Market Impact and Economic Scale

Mechanical filtration is a fundamental component of the global water filtration market, which is valued at over $30 billion. This value reflects its widespread application and necessity.

Growth and Innovation: The market is continuously growing, with innovations focused on improving efficiency and lifespan of filters. Developments in materials science, such as the use of advanced polymers and nano-materials, are enhancing the capabilities of mechanical filters.

Global Demand: The demand for mechanical filtration is high across the world, especially in regions facing water scarcity and pollution. Its role in pretreatment makes it indispensable in comprehensive water purification setups.

Industry Integration: It's a cornerstone technology in industries beyond just water treatment, including pharmaceuticals, food and beverage production, and chemical processing. Mechanical filtration, despite its limitations, remains a crucial and foundational element in water purification and treatment processes worldwide. Its simplicity, cost-effectiveness, and efficiency make it an indispensable first step in most water treatment systems. The market for mechanical filtration reflects its universal applicability and the continuous need for water purification across various sectors.

ACTIVATED CARBON FILTERS

Applications

Home Water: widely used in residential water filtration systems to improve the taste and quality of drinking water.

Industrial Applications: employed in various industries to treat water and remove organic compounds, chlorinated compounds, and gases.

Municipal Water Treatment: used in large-scale water treatment facilities to improve the palatability of drinking water.

Air Purification: besides water, activated carbon filters are also used in air purifiers to remove odours and harmful compounds.

Advantages

Taste and Odour Removal: exceptionally effective in making water more palatable by removing chlorine and other compounds that give water an unpleasant taste or smell.

Organic Contaminant Removal: efficient in adsorbing and removing a wide range of organic compounds, including some pesticides and solvents.

Versatility: can be used in various forms such as granular, block, or impregnated carbon, catering to different filtering needs.

Chemical Adsorption: offers a unique advantage in removing certain chemicals that are challenging to eliminate through other means.

Limitations

Ineffectiveness Against Inorganics: does not remove inorganic substances like salts, minerals, and metals.

Limited Microbial Action: cannot remove bacteria, viruses, or other microorganisms.

Saturation: the filter becomes less effective over time as the activated carbon becomes saturated with adsorbed contaminants.

Maintenance: requires regular replacement to maintain effectiveness.

Market Impact and Economic Scale

The global market for activated carbon filters is substantial, with a valuation of approximately $4.7 billion, and is expected to grow, reflecting their widespread use.

Innovation and Research: there is ongoing research into enhancing the adsorption capacity and efficiency of activated carbon, including the development of modified or treated carbon materials.

Environmental Concerns: demand for these filters is partly driven by increasing environmental awareness and the need for cleaner water in both residential and industrial contexts.

Diverse Applications: beyond water purification, activated carbon filters find roles in various sectors, including wastewater treatment, air purification, and even in protective gear, contributing to its broad market impact.

Activated carbon filters are a cornerstone in modern

water purification, prized for their ability to improve water taste and remove specific contaminants. Their versatility and effectiveness in adsorbing organic compounds make them a popular choice in both domestic and industrial applications. The market for these filters is significant and continues to expand, driven by ongoing environmental concerns and the quest for cleaner water.

REVERSE OSMOSIS (RO)

Applications

Residential Use: common in household water purification systems for drinking water, effectively removing a wide array of contaminants.

Commercial Settings: used in restaurants, hotels, and offices for providing high-quality water for drinking and cooking.

Industrial Applications: essential in sectors like pharmaceuticals, food and beverage manufacturing, and power plants for process water treatment.

Desalination: widely employed in desalination plants across coastal regions for converting seawater into potable water.

Advantages

Broad Range Contaminant Removal: exceptionally effective in removing a wide variety of contaminants, including salts, minerals, bacteria, viruses, and organic compounds.

Improved Water Quality: delivers high-quality water, often surpassing the standards set for drinking water.

Versatility: can be scaled and adapted for various applications, from small under-the-sink units to large municipal treatment plants.

Selective Permeability: the semi-permeable membrane allows selective removal of contaminants while retaining essential minerals in water.

Limitations

Energy Intensive: among the more energy-consuming purification methods, especially in large-scale applications like desalination.

Water Waste: generates a significant amount of brine waste, which poses environmental disposal challenges.

Maintenance: requires regular maintenance and membrane replacement.

Pre-Treatment Needs: water often needs pre-treatment to prevent membrane fouling and damage.

Market Impact and Economic Scale

The global Reverse Osmosis (RO) market is valued at around $4 billion, underlining its significance in the water treatment industry.

Growing Demand: there is an increasing demand for RO systems, driven by worsening water scarcity and quality concerns globally.

Technological Advancements: continuous innovations are focused on reducing energy consumption and improving membrane technology, making RO systems more efficient and sustainable.

Desalination Dominance: a major player in the desalination industry, RO technology is central to addressing water scarcity in arid regions and islands.

Reverse Osmosis stands out as one of the most effective water purification methods available today. Its ability to remove a wide range of contaminants makes it a preferred choice in various settings, from residential to industrial. The technology's market impact is substantial and continues to grow, fueled by global demands for clean and safe drinking water and advancements in RO technology, making it more efficient and environmentally friendly.

DISTILLATION

Applications

Laboratory Use: widely used in scientific research settings for obtaining high-purity water, essential for various experiments and analyses.

Medical Facilities: employed in hospitals and clinics for applications where ultra-pure water is necessary.

Residential Purification: some households use distillation units for drinking water purification, particularly in areas with heavily contaminated water sources.

Industrial Processes: in industries such as pharmaceuticals and food processing, where water of exceptional purity is required.

Advantages

High Purity: capable of removing a wide array of contaminants, including bacteria, viruses, heavy metals, and minerals, to produce very pure water.

No Chemicals Used: distillation doesn't rely on chemical treatment, making the water safe from chemical contamination.

Effectiveness Against Hard-to-Remove Contaminants: effective in removing contaminants that are difficult to eliminate through other purification methods.

Simplicity of Process: the process of distillation is straightforward, involving evaporation and condensation, making the technology easy to understand and implement.

Limitations

Energy Intensive: one of the most energy-consuming water purification methods, making it less practical for large-scale applications.

Slow Process: the rate of water production is generally slow, which can be a limitation for high-demand applications.

Does Not Remove Certain Chemicals: volatile organic compounds with a lower boiling point than water may not be effectively removed.

Maintenance and Cost: requires regular cleaning to remove scale and residue; the initial setup and operation costs can be high.

Market Impact

Niche Market Presence: primarily focused on specialized sectors such as laboratories and medical facilities, contributing to a specific segment of the larger water purification market.

Growth in Specific Applications: the market for distillation units is stable, with growth primarily in sectors where high-purity water is a critical requirement.

Technological Developments: innovations focus on making distillation units more energy-efficient and capable of higher output, which could expand their market presence.

Environmental Concerns: the high energy requirement is a constraint, especially in the context of increasing environmental sustainability concerns.

Distillation is a time-tested method, valued for its ability to produce highly pure water. While its application is limited by high energy requirements and slow processing rate, it remains indispensable in settings where the highest water purity is paramount. The market for distillation units, though niche compared to other technologies, is significant in sectors where no other purification method can achieve the required water quality. Advances in technology that address its energy consumption and efficiency could widen its applicability and market presence.

ULTRAVIOLET (UV) DISINFECTION

Applications

Residential Water Treatment: commonly used in home water purification systems, especially as a final disinfection step.

Municipal Water Treatment: employed at the municipal level for disinfecting large volumes of water, often as part of a multi-stage treatment process.

Commercial Settings: used in various commercial establishments like hotels, resorts, and restaurants for ensuring water safety.

Aquariums and Pools: effective in

controlling microbial growth in recreational water and aquariums.

Advantages

Effective Pathogen Control: highly effective against a broad range of pathogens, including bacteria, viruses, and protozoa.

Chemical-Free: no chemicals are involved in the process, avoiding the risk of chemical contamination.

Rapid Treatment: ultraviolet disinfection occurs quickly compared to some chemical methods.

Low Maintenance: UV systems generally require less maintenance than other disinfection methods, mainly needing periodic bulb replacement.

Limitations

No Residual Disinfection: unlike chemical methods, UV provides no lasting protection against recontamination.

Water Clarity Dependency: the effectiveness of ultraviolet light is hindered by turbid or coloured water, which can block or absorb the UV rays.

Ineffective Against Non-Microbial Contaminants: does not remove chemical contaminants, heavy metals, or dissolved solids.

Energy Consumption: while generally less energy-intensive than methods like distillation, UV systems do require a constant power source.

Market Impact and Economic Scale

Growing Market: the global UV disinfection equipment market is valued at approximately $1.3 billion, with expectations for continued growth.

Diverse Applications: beyond water treatment, ultraviolet disinfection is gaining traction in air and surface disinfection, broadening its market.

Technological Innovation: ongoing advancements are focused on increasing efficiency, reducing energy consumption, and extending the lifespan of UV lamps.

Environmental and Health Awareness: increasing awareness about the health risks associated with chemical disinfectants is driving the market for UV disinfection solutions.

Ultraviolet disinfection is a popular choice in various settings for its effectiveness in pathogen control and chemical-free process. Its application in residential, municipal, and commercial water treatment highlights its versatility. The technology's market is expanding, fueled by technological innovations and a growing preference for non-chemical disinfection methods. As concerns over chemical contaminants in water and the environment grow, UV disinfection stands out as a preferred solution, contributing to its significant market presence and potential for future growth.

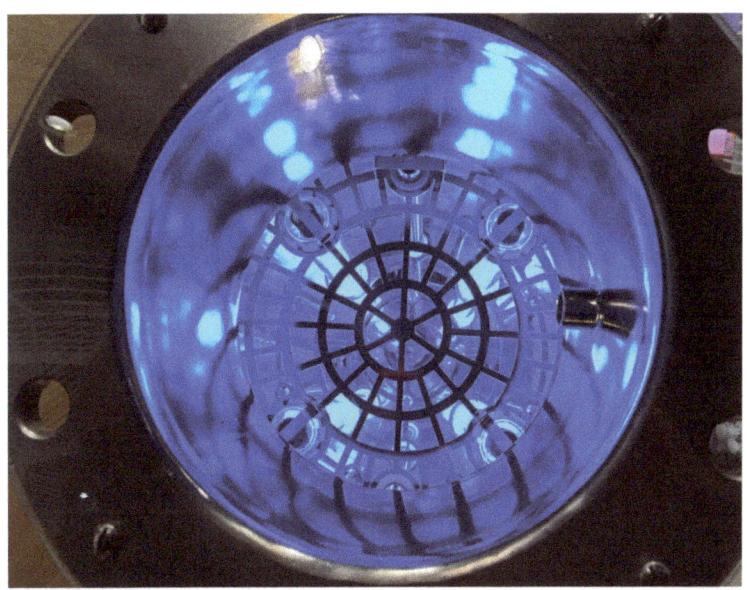

ION EXCHANGE

Applications

Residential and Commercial Settings: used in water softening to remove calcium and magnesium.

Industrial Processes: employed to purify water and recover valuable minerals.

Demineralization Processes: integral in power plants and large industrial applications.

Medical Field: used for producing ultra-pure water for dialysis.

Advantages

Highly effective for softening hard water and removing specific ions.

Essential for applications requiring demineralized or deionized water.

Can be tailored to target specific contaminants.

Limitations

Not effective against organic contaminants, viruses, or bacteria.

Requires regular regeneration and maintenance.

Inefficient for non-ionized particles and dissolved organic substances.

Market Impact

Ion exchange is significant in industrial applications, contributing to the broader water treatment chemicals market.

There is constant demand in sectors where water

hardness and specific ion removal are critical.

Innovations focus on more efficient resins and sustainable regeneration methods.

SEDIMENTATION AND CLARIFICATION

Applications

Municipal Water Treatment Plants: primary step for removing suspended solids.

Industrial Use: various industries use it for preliminary water treatment and wastewater management.

Stormwater Management: integral in stormwater management and treatment processes.

Advantages

Simple and cost-effective method for reducing water turbidity.

Enhances the effectiveness of subsequent treatment processes.

Limitations

Limited to removing particulate matter; does not affect dissolved substances.

Requires large settling tanks and space, particularly in municipal settings.

Market Impact

A foundational part of the water treatment infrastructure, contributing to the overall market for water treatment equipment.

Continuous demand in urban and industrial development, with innovations focusing on improving efficiency and reducing footprint.

CERAMIC FILTRATION

Applications

Portable Water Filters: popular for outdoor activities and emergency situations.

Household Use: particularly in regions with unreliable water supply.

Industrial Applications: employed in certain industries for fine filtration.

Advantages

Effective against bacteria and protozoa.

Durable and long-lasting with proper maintenance.

Does not require electricity, suitable for remote locations.

Limitations

Ineffective against viruses and chemical contaminants.

Flow rate can be slow, and filters may clog, requiring regular cleaning.

Market Impact

Significant in the portable water filter market and emergency preparedness sector.

Growing demand in areas with limited access to safe drinking water.

Ongoing improvements aim to enhance efficiency and flow rates.

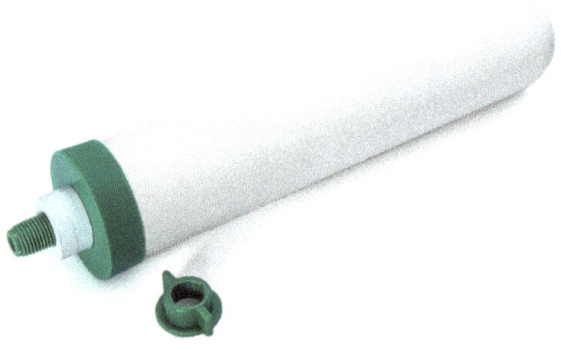

OZONE WATER TREATMENT

Applications

Municipal Water Treatment: widely used for effective disinfection.

Industrial Uses: Employed in bottled water industries and swimming pools for oxidation and disinfection. Used in industrial wastewater treatment to degrade organic pollutants.

Advantages

Powerful oxidizing agent, highly effective against a wide range of pathogens.

Leaves no chemical residue in water.

Limitations

Does not remove dissolved solids or heavy metals.

Requires careful handling due to its highly reactive nature.

Market Impact

Growing presence in industrial wastewater treatment and bottled water sectors.

Innovations focus on improving efficiency and integrating ozone generation with other treatment processes.

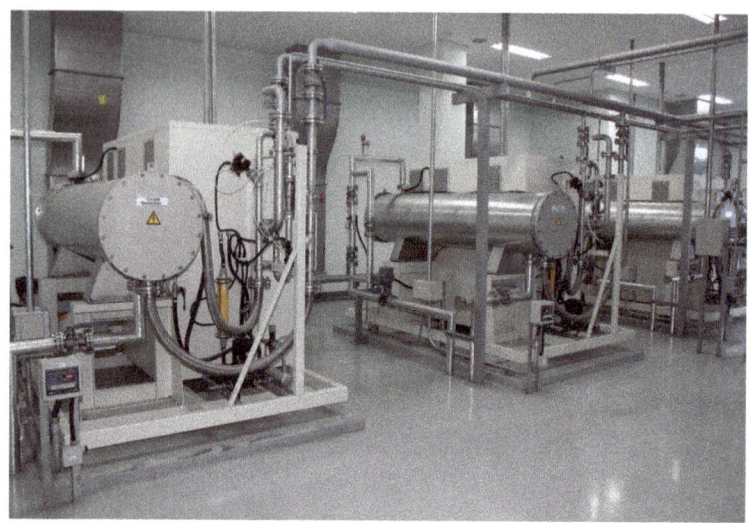

BIOLOGICAL FILTRATION

Applications

Sewage Treatment Plants: used for organic and nutrient removal.

Aquaculture and Aquarium Systems: employed for maintaining water quality.

Water Purification Systems: integral in certain natural water purification systems.

Advantages

Environmentally friendly, using natural biological processes.

Effective for organic matter and nutrient removal.

Limitations

Slow process, not suitable for removing all types of contaminants.

Requires careful monitoring and maintenance.

Market Impact

Mainly used in municipal wastewater treatment, contributing to the larger wastewater management industry.

Growing interest in incorporating natural and sustainable treatment methods.

ELECTRODEIONIZATION (EDI)

Applications

Industrial Applications: Used for producing ultra-pure water, such as in pharmaceuticals and electronics manufacturing. Employed in power plants and laboratories for water purification.

Advantages

Continuous process that combines ion exchange and electrodialysis.

Chemical-free and produces high-purity water.

Limitations

Requires feed water that is already partially purified.

Not effective against non-ionic contaminants.

Market Impact

Niche but important in sectors requiring high-purity water.

Innovations focus on increasing efficiency and reducing operational costs.

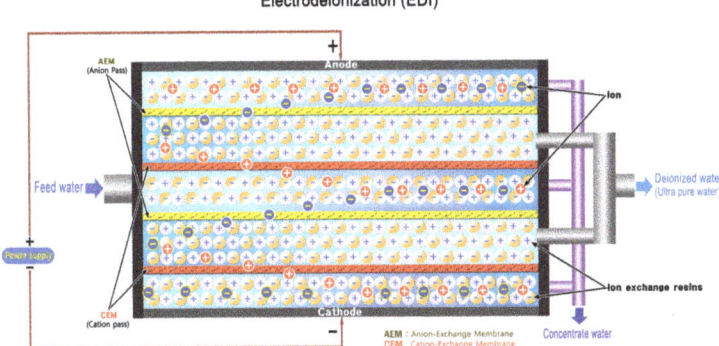

Electrodeionization (EDI)

NANOTECHNOLOGY

Applications

Pollutant Removal: emerging use in removing pollutants like heavy metals, microorganisms, and organic compounds from water.

Wastewater Treatment: potential applications in industrial wastewater treatment for targeting specific contaminants.

Advantages

High efficiency in removing contaminants at the molecular level.

Potential for selective targeting and removal of specific pollutants.

Innovations in nanotechnology are leading to more efficient and effective water purification methods.

Limitations

Concerns about the release of nanoparticles into the environment and potential health risks.

High cost of development and implementation.

Challenges in scaling up the technology for widespread use.

Market Impact

Rapidly growing segment within the water treatment industry, with significant potential for expansion.

Research focus on developing nanofiltration membranes and nanoparticle-based purification.

Ongoing research and development are expected to drive market growth, with a focus on sustainability and efficiency.

The potential for significant breakthroughs in water purification technology makes it a highly watched and invested sector.

FORWARD OSMOSIS

Applications

Desalination and Wastewater Treatment: increasing interest for use in these areas.

Food Processing: applications in food processing for concentration processes without heat.

Advantages

Lower energy requirements compared to reverse osmosis.

Effective in treating high-salinity water and some types of industrial wastewater.

Limitations

Slower process and requires a draw solution, which needs to be regenerated.

Challenges in membrane fouling and scaling.

Market Impact

Still a developing market but with growing interest due to its potential in sustainable water treatment.

Research focus on integrating forward osmosis with other treatment technologies.

Research and pilot projects are expanding, indicating a future increase in market presence, especially in regions where energy efficiency is crucial.

ULTRASONIC WATER TREATMENT

Applications

Experimental stage for controlling algae growth in water bodies and microbial control.

Potential use in industrial wastewater treatment for reducing pollutant load.

Advantages

Chemical-free process, reducing the risk of chemical contamination.

Effective against certain types of bacteria and algae.

Limitations

Limited research on overall effectiveness and scalability.

Energy-intensive, which can limit its practical application.

Market Impact

Currently a niche market with potential growth in industrial water treatment.

Continued research and technological improvements could expand its applications and market size.

ELECTRODIALYSIS

Applications

Primarily used for desalination, especially of brackish water.

Applied in the food industry for salt removal and concentration of certain products.

Advantages

Selectively removes ions, making it effective for specific types of saltwater.

More energy-efficient than reverse osmosis for certain applications.

Limitations

Ineffective against non-ionic contaminants.

Membranes can be susceptible to fouling and scaling.

Market Impact

Part of the broader desalination market, which is growing due to increasing water scarcity.

Continued improvements and innovations in membrane technology could increase its efficiency and reduce costs, potentially expanding its market.

Each of these technologies plays a vital role in addressing specific water purification needs, with their market impact reflecting their effectiveness, applicability, and the ongoing technological advancements in the field.

Case Studies

The implementation of reverse osmosis in desalination plants in arid regions like the Middle East has revolutionized water accessibility. UV purification systems have been effectively used in rural communities and emergency relief operations, providing clean water without extensive infrastructure.

Impact Assessment: These modern technologies have significantly improved water quality and accessibility. They have also contributed to public health improvements in regions where waterborne diseases were prevalent.

Limitations and Technical Challenges: Despite their effectiveness, these modern technologies face challenges. Reverse osmosis systems require high energy and are costly to maintain. UV purification systems are less effective in turbid water. The disposal of contaminants, especially from desalination processes, poses environmental concerns.

Environmental Impact: The environmental footprint of water purification methods is a growing concern. Chemical methods can result in harmful by-products. The energy requirements of advanced technologies contribute to carbon emissions, and the disposal of brine from desalination impacts marine ecosystems.

Economic and Accessibility Issues: The cost of implementing modern technologies is prohibitive for many regions, especially in developing countries. Accessibility remains a challenge where infrastructure and economic resources are limited.

My Insights on the Efficacy of Current Solutions

Throughout my research, I have observed that the choice of water purification method greatly depends on the context—geographic, economic, and social. In some cases, traditional methods are more feasible, while in others, modern technologies are essential.

In comparing the various techniques, it becomes clear that no single solution fits all scenarios. Each process has its unique application and suitability based on the specific requirements and constraints of the region.

Looking ahead, the focus should be on developing sustainable, energy-efficient, and cost-effective purification methods. Innovations should aim to reduce environmental impact while enhancing accessibility and affordability.

In conclusion, while significant advancements have been made in water purification, challenges in sustainability, efficiency, and accessibility persist.

My journey in exploring these methods has reinforced my belief in the need for holistic solutions that consider environmental, economic, and social dimensions. I advocate for a collaborative approach where governments, private sectors, and communities work together to innovate and implement sustainable water purification solutions.

Footnotes:

(1): Cotruvo, J. A. (2017). "Water Purification: Historical Perspectives and Modern Technologies." Journal of Water and Health.

(2): Crittenden, J. C., Trussell, R. R., Hand, D. W., Howe, K. J., & Tchobanoglous, G. (2012). MWH's Water Treatment: Principles and Design.

(3): Dąbrowska, L., & Nawrocki, J. (2014). "Application of activated carbon for water and wastewater treatment." In L. Pawłowski, M. Dudzińska, & A. Pawłowski (Eds.), Environmental Engineering IV (pp. 21–33).

(4): Greenlee, L. F., Lawler, D. F., Freeman, B. D., Marrot, B., & Moulin, P. (2009). "Reverse osmosis desalination: Water sources, technology, and today's challenges."

CHAPTER 5: ILLUMINATING PURITY: UV TECHNOLOGY IN WATER TREATMENT

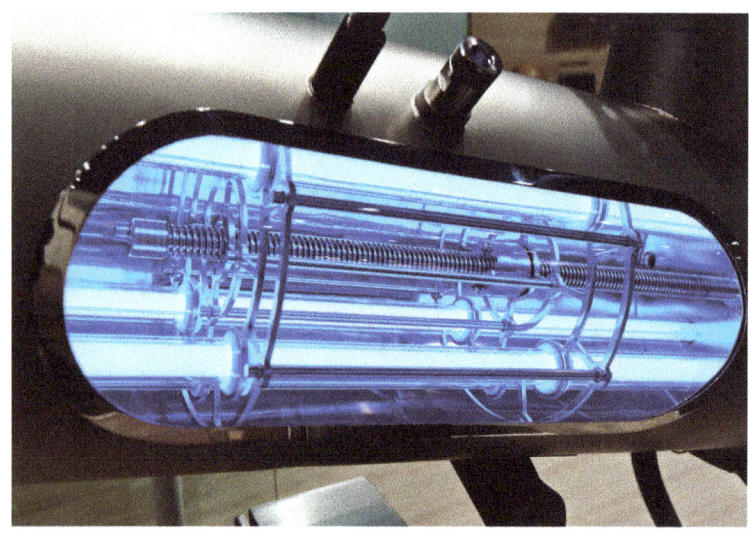

Imagine walking the streets of Marseille in 1910, witnessing one of the earliest uses of ultraviolet light in water purification. This pioneering step was a game changer. It's fascinating to think how this innovation spread from this French city to the rest of the world, significantly reducing waterborne diseases. It reminds me of the power of a single breakthrough to change the course of public health.

The History of UV Light as a Purification Method for Waterborne Diseases

Before delving into the technical aspects of UV technology, it's important to understand its historical context. The use of ultraviolet light as a method for purifying water dates back to the early twentieth century. The discovery that it could inactivate microorganisms led to its first practical applications in water treatment. In 1910, Marseille, France, saw one of the earliest implementations of UV water purification. This system used UV light to treat the city's water supply, significantly reducing the prevalence of waterborne diseases such as cholera and typhoid fever. (1)

The success in Marseille paved the way for further adoption of UV purification, particularly in Europe and North America. Throughout the twentieth century, as the understanding of UV light's germicidal properties expanded, so did its applications in water treatment. Advances in UV lamp technology during the mid-century enhanced the effectiveness and affordability of ultraviolet systems, leading to more widespread use. By the late twentieth and early twenty-first centuries, UV water purification had become a recognized and valued method for ensuring safe drinking water, both in municipal water supplies and in individual and commercial settings.

Understanding UV Light

I remember first learning about UV light—it felt like discovering invisible magic. UV-C, in particular, is a remarkable part of the spectrum. Invisible yet powerful, it can break down the molecular bonds of harmful microorganisms. I often think about how something so small and unseen can have such a significant impact on our health.

Ultraviolet light, a component of the electromagnetic spectrum, is a type of energy invisible to the human eye. It is categorized into three primary types based on wavelength: UV-A, UV-B, and UV-C.

UV-A (320–400 nm) and UV-B (280–320 nm) are commonly known for their roles in causing suntans and sunburns, respectively. UV-C (100–280 nm), with the shortest wavelength, possesses the highest energy and is particularly significant in water purification. UV-C light's properties are unique; it is not only invisible but also carries enough energy to break chemical bonds. (2) This makes it highly effective against various microorganisms, including bacteria, viruses, and protozoa. Unlike UV-A and UV-B, which reach the earth's surface, UV-C is absorbed by the ozone layer, making it rare in nature but can be artificially generated for water treatment purposes.

Mechanisms of Action

It's incredible how UV-C light, by creating pyrimidine dimers in DNA and RNA, can stop harmful pathogens right in their tracks. It's akin to placing a barrier on a road; the pathogens simply can't replicate or infect. This aspect of UV technology always reminds me of the intricate dance of nature and science.

UV-C light inactivates microorganisms by disrupting their DNA and RNA. When microorganisms are exposed to UV-C, the light penetrates their cells and is absorbed by the nucleic acids. This absorption leads to the formation of new bonds between adjacent nucleotides, creating photoproducts such as pyrimidine dimers. (3) These dimers effectively prevent the DNA and RNA strands from unzipping and replicating. Without the ability to replicate, the microorganisms cannot infect or reproduce, rendering them harmless.

UV purification stands out as a safe and environmentally friendly method of water treatment. Unlike chemical disinfectants, ultraviolet light leaves no residual chemicals in the water. This absence of residuals means that there is no risk of overtreatment or the creation of harmful by-products, which can be a concern with methods like chlorination. Additionally, UV purification does not alter the taste, odour, or pH of the water, maintaining its natural quality.

From an environmental standpoint, UV purification is a sustainable choice. It does not require the transport or storage of hazardous chemicals, reducing the risk of spills and chemical accidents. The process is energy efficient, especially with the advent of advanced UV-C LEDs, making it a more sustainable option compared to energy-intensive methods like boiling or distillation.

The Future is UV-C LED

One of the most interesting developments that I believe will have a significant future impact is the UV-C LED light, a technology referred to as "SteriLight." SteriLight represents a major advancement in disinfection and sterilization processes, operating in the Ultraviolet-C spectrum. (4) Its effectiveness in inactivating harmful microorganisms through DNA and RNA disruption positions it as a powerful tool in promoting environmental and public health.

SteriLight's applications are diverse, extending to water treatment, surface and sterilization. Despite facing challenges like low external quantum efficiency and thermal management issues, its market is rapidly expanding. From $20 million in 2008 to an expected $991 million by 2023, SteriLight is carving a significant niche in the LED market.

The technology's compact size and energy efficiency, coupled with its mercury-free nature, make it an environmentally friendly and versatile option compared to traditional mercury lamps. Its growing adoption in various sectors, especially in water purification and medical sterilization, underscores its potential to revolutionize disinfection practices globally. However, the development of comprehensive safety standards and addressing technical challenges will be crucial for its continued growth and wider application.

UV-C LED technology represents a significant advancement in the field of disinfection and sterilization, offering various applications in water, surface, and air disinfection. This technology operates in the Ultraviolet-C spectrum, specifically emitting radiation in the range of 180nm to 280nm wavelengths. This range is particularly effective in inactivating microorganisms by damaging their DNA and RNA, thus preventing them from reproducing and spreading infections.

Water Disinfection: UV-C LEDs are revolutionizing water treatment processes by providing a highly efficient method for removing, deactivating, or killing pathogenic microorganisms in water. They have been integrated into systems of varying scales, from small domestic settings to industrial-scale systems.

Surface Disinfection: In settings like food production lines, UV-C LEDs ensure rapid and reliable surface disinfection. They have been integrated into high-touch areas such as escalator belts and door handles, to minimize infection spread.

Air Disinfection: in crowded public spaces like airports and hospitals, UV-C LED technology is employed to disinfect the air, proving effective against airborne microorganisms like viruses, bacteria, yeasts, and mould.

Technical Aspects and Benefits: UV-C LEDs emit light at specific, narrow wavelengths, with 265nm being the most effective for disinfection.

They are solid-state devices, different from traditional sources that use heated filaments or gas discharge. They offer several benefits over conventional mercury lamps, including being mercury-free, having longer lifetimes, enduring numerous on/off cycles, and offering instant on/off capability.

UV-C LEDs are energy-efficient, producing less heat and requiring less energy during operation, and their compact size and rugged build make them versatile for various applications.

Market Size and Future Potential: The UV-C market has shown significant growth, increasing from $20 million in 2008 to $144 million in 2019. It is projected to reach $991 million by 2023, indicating a strong potential for UV-C LED technology in various applications.

This growth is driven by the increasing demand for water purification due to the scarcity of clean water resources, and the need for sterilization in medical applications.

Challenges and Limitations: UV-C LEDs face technical challenges such as low external quantum efficiency (EQE) of the chip, aging of the packaging material, and thermal management issues.

While both UV-C LEDs and mercury lamps have been proven effective for disinfection, mercury lamps currently have better optical power and a lower price, making them more widely adopted.

Safety is a crucial consideration, as UV-C light can cause damage to human skin and eyes. Safety standards for UV-C LEDs are still developing, particularly as they gain market share in consumer products.

I believe that UV-C LED technology, with its advantages and burgeoning market size, represents a significant step forward in environmental and public health protection. However, addressing the technical and safety challenges will be crucial for its continued growth and wider adoption.

UV-C Versus Traditional Methods

I've always been concerned about the downsides of traditional methods like chlorination—the chemical by-products, for one. Ultraviolet purification, by contrast, feels cleaner and safer. It's like choosing a path that's not only effective but also responsible.

When comparing UV purification with traditional methods such as chlorination and boiling, several key differences become apparent. Chlorination, a widely used method, involves adding chlorine or chlorine compounds to water to kill bacteria and viruses. While effective in disinfection, chlorination can lead to the formation of harmful by-products like trihalomethanes (THMs) and haloacetic acids (HAAs), which are linked to health risks like cancer and reproductive issues.

Boiling water is another common method, especially in household settings. It is effective against most pathogens; however, boiling is energy-intensive and impractical on a large scale. It also does nothing to remove chemical contaminants and requires a reliable energy source, which might not be available in all settings.

Advantages of UV in Terms of Efficiency and Environmental Impact

UV purification offers several advantages over these traditional methods. In terms of efficiency, ultraviolet systems can process large volumes of water quickly without altering the water's taste or chemical composition. They are also effective against a broader range of microorganisms, including chlorine-resistant protozoa like Cryptosporidium and Giardia.

The environmental impact of UV purification is significantly lower compared to chlorination and boiling. UV systems do not produce harmful chemical by-products and have a smaller carbon footprint. Their energy efficiency, particularly with the advent of UV-C LED technology, further enhances their appeal as a sustainable water treatment solution.

From Theory to Practice: Applying UV Technology in Water Purification

The journey of UV technology from theory to practice is a testament to human ingenuity. I've seen small UV units make a huge difference in remote communities. It's a vivid reminder of how technology can be tailored to meet diverse needs.

As the understanding of ultraviolet technology's effectiveness in water purification has grown, its transition from theoretical concepts to practical applications has accelerated. This section will delve into the practical implementation of UV systems, discussing their design, adaptability, and real-world usage in various contexts, from small community setups to large-scale municipal plants.

Design Considerations: Designing an effective UV purification system involves considerations such as the water's optical properties, flow rate, and the specific pathogens targeted for inactivation. System design must ensure that all water passing through the reactor receives adequate UV exposure, a challenge met through precise engineering and innovative reactor designs.

Adapting UV Systems for Diverse Applications: UV technology's versatility allows for its adaptation in various settings. Portable ultraviolet units have become popular for individual and emergency use, while larger systems are being integrated into existing municipal treatment plants. This adaptability extends to industries such as pharmaceuticals and food processing, where water purity is paramount.

Designing UV Systems: Designing a UV system is a complex challenge. It's about finding the perfect balance of light intensity, water flow, and exposure. Every time I see a well-designed system, I am reminded of the harmony between technology and nature. The engineering of UV water purification systems is a complex process that requires a careful balance between several key factors. The design must account for the intensity of ultraviolet light, the exposure time, and the water flow rate to ensure all microorganisms are effectively inactivated. System designers also consider the water's optical properties, including clarity and UV transmittance, which can significantly affect the UV dose required for effective treatment.

Footnotes:

(1): Kowalski, Wladyslaw. "Ultraviolet Germicidal Irradiation Handbook: UVGI for Air and Surface Disinfection." Springer, 2009.

(2): Bolton, James R., and Christine A. Cotton. "The Ultraviolet Disinfection Handbook." American Water Works Association, 2008.

(3): Hijnen, W.A.M., Beerendonk, E.F., & Medema, G.J. "Inactivation credit of UV radiation for viruses, bacteria and protozoan (oo)cysts in water: A review."

(4): Lawal, Oluwaseun, et al. "UVC LED Irradiation for Disinfection of Waterborne Bacteria Escherichia Coli, Salmonella Enterica, and Staphylococcus Aureus."

CHAPTER 6: THE BIOFOULING GLOBAL CHALLENGE: OUR PIPES ARE THE PROBLEM

 In my journey to understand how can we improve our water systems, I've encountered many challenges, but one of the most significant and often overlooked is biofouling. This phenomenon, where microorganisms and small pathogens accumulate on wet surfaces, particularly within our water pipes, poses a serious threat to the integrity and efficiency of our water delivery systems, and consequently, to public health.

When I think about the problem of biofouling, it's really a testament to how resilient and adaptable biological organisms are. At the same time, it highlights the vulnerabilities in how we've designed and maintained our water systems. Our pipes, essentially the arteries of our water systems, are at the center of this issue. The materials they're made from, their age, and their design all play crucial roles in how severely biofouling affects them. Over time, these pipes can become hotspots for unwanted biological growth, leading to reduced efficiency, increased maintenance costs, and most alarmingly, a decline in our water quality.

Having visited numerous demolition sites, I've seen firsthand the inside of old pipes. They're often coated with thick layers of slimy, foul-smelling biofilm, evidence of years of biological buildup. These visual reminders the critical need to address biofouling in our water systems.

Globally, the challenge of biofouling is as diverse as it is widespread, affecting both developed and developing regions. From small rural communities to sprawling urban centres, the implications of bio-fouling in water systems are profound, impacting millions of people and the ecosystems surrounding them. The issue transcends mere inconvenience, posing significant health risks, as biofouling can harbour harmful pathogens and facilitate the spread of waterborne diseases.

This chapter will explore the intricate nature of biofouling, dissecting how and why our pipes become compromised. We will explore the mechanics of biofouling, understanding its stages and the factors that exacerbate its growth. By examining global case studies, we will gain insights into the real-world impact of biofouling on water systems and the varied responses to this challenge. Furthermore, we will investigate the role of pipe materials and design in promoting or mitigating biofouling, scrutinizing the flaws in current infrastructures and the potential for innovative alternatives. The chapter will culminate in a discussion on strategies for mitigation and prevention, highlighting cutting-edge technologies, effective policy approaches, and best practices in water system management.

The issue of biofouling is a call to action, urging us to rethink how we design, maintain, and renew our water systems. It challenges us to innovate and adapt, ensuring that the lifelines of our communities—the pipes that carry our most vital resource—are not only efficient but also safe. As we navigate through this chapter, we will uncover the complexities of biofouling and the critical steps needed to turn this global challenge into an opportunity for advancement in water system management.

Definition and Overview

Biofouling refers to the undesirable accumulation of microorganisms, plants, algae, and small animals on wetted surfaces, particularly in water systems. This phenomenon is not just a minor nuisance; it represents a significant global concern for several reasons.

Firstly, biofouling can lead to the degradation of water quality, introducing harmful bacteria and other pathogens into the water supply. Secondly, it reduces the efficiency of water systems by clogging pipes and reducing flow, which can lead to increased energy costs and the need for more frequent maintenance. Finally, in severe cases, biofouling can cause complete system failures, posing substantial risks to public health and safety.

The organisms involved in biofouling vary widely but can be broadly categorized into microorganisms (like bacteria and fungi), plants (such as algae), and small animals (including mollusks and barnacles). Each of these groups contributes differently to biofouling. Bacteria, for example, often form slimy layers known as biofilms, which can adhere strongly to surfaces and act as a base for other organisms to attach. Algae, on the other hand, can clog systems and reduce light penetration, affecting water treatment processes. Small animals like barnacles and mollusks can physically obstruct water flow and damage infrastructure. (1)

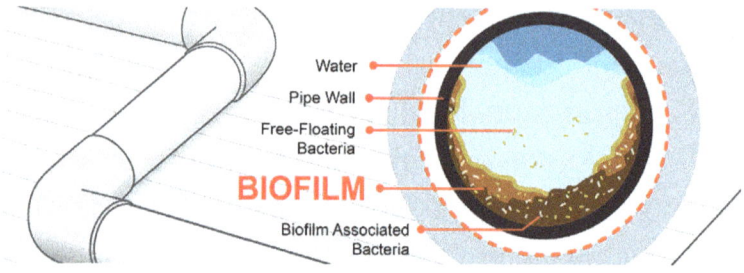

Mechanics of Biofouling

Biofouling in water systems, particularly in pipes, occurs in several stages. The initial stage involves the adhesion of microorganisms to the pipe surface. These organisms secrete extracellular polymeric substances, forming a sticky matrix that adheres to the pipe and provides a scaffold for further colonization. This stage is critical, as it sets the foundation for more complex biofouling communities.

The next stage involves the growth and maturation of these biofilms. During this phase, the biofilms become more diverse as different species of bacteria, algae, and small organisms colonize the surface. This diversity creates a more complex and robust biofouling layer that is harder to remove and can start to affect water flow and quality.

The final stage of biofouling is characterized by the accumulation of larger organisms like mollusks and barnacles. This stage significantly impacts the mechanical integrity and function of water systems.

The weight and size of these organisms can cause physical blockages in pipes and can even lead to corrosion and structural damage.

Each stage of biofouling development presents unique challenges and requires specific strategies for mitigation and control. Understanding these stages is crucial for effective management of water systems and maintaining the quality and efficiency of water delivery.

Impact of Biofouling on Water Systems

Reduced Efficiency and Flow:

Biofouling significantly impacts the efficiency of water systems. One of the primary effects of biofouling is the reduction in the diameter of pipes due to the buildup of organisms and materials on their inner surfaces. This phenomenon, known as "pipe clogging," can drastically reduce the flow rate of water through the system. The accumulation of biofouling layers creates resistance to water flow, necessitating increased pressure and energy to maintain desired water delivery rates. This elevation in operational costs is accompanied by additional strain on mechanical components, leading to more frequent breakdowns and repairs.

Moreover, biofouling can lead to uneven water distribution within the system, resulting in areas of stagnation, where water movement is minimal. Stagnant water exacerbates biofouling and poses a risk for the proliferation of harmful microorganisms. The implications for water pressure and flow rates are substantial, affecting everything from residential water supply to industrial and municipal water delivery systems. The environmental impact is also notable, as the increased energy required for water treatment and pumping contributes to the carbon footprint of water utilities. (2)

Water Quality Degradation:

The impact of biofouling on water quality is a major concern. Biofilms, a key component of biofouling, can harbour a range of pathogens, including harmful bacteria such as Legionella and E. coli, along with protozoa and viruses. These microorganisms, protected within the biofilm structure, are often more resistant to standard water treatment methods, posing significant risks to human health. The contamination of drinking water supplies with these pathogens can lead to various illnesses, ranging from gastrointestinal infections to more severe diseases.

Biofouling also alters water chemistry, affecting pH levels, dissolved oxygen, and nutrient concentrations. These changes can further degrade water quality, rendering it unsuitable for consumption or specific uses. The health risks associated with contaminated water are particularly acute for vulnerable populations, including the elderly, children, and those with weakened immune systems.

Additionally, the presence of biofouling complicates water treatment processes. Ensuring water meets health and safety standards in the face of biofouling may require more intensive treatment steps, increased chemical use, and stringent monitoring. This not only raises the cost and complexity of water treatment but also poses questions about the sustainability and environmental impacts of these practices. (3)

Pipes: The Heart of the Problem

Material and Design Flaws:

The materials that pipes are made with play a crucial role in the susceptibility of water systems to biofouling. Traditional pipe materials like iron, steel, and concrete have surface properties that can promote the adhesion and growth of biofouling organisms. These materials often have microscopic roughness and porosity, providing ideal niches for microorganisms to anchor and form biofilms. Additionally, certain materials may react chemically with water, creating conditions that further support biofouling.

The design of pipes also influences biofouling. Pipes with complex geometries, sharp bends, and rough internal surfaces offer more areas where water flow is disrupted, creating pockets of stagnation that are conducive to biofilm formation. Moreover, joints and seams in pipe systems can become focal points for biofouling, as they may allow for minor leaks or water seepage, further encouraging microbial growth.

In response, there have been advancements in pipe materials and design aimed at mitigating biofouling. Materials like copper and certain plastics have smoother surfaces and are less reactive chemically, making them less prone to biofilm formation. Pipes designed with fewer joints and smoother bends can reduce areas of stagnation. Moreover, coatings and linings that are anti-microbial or smoother can be applied to the interior surfaces of pipes, reducing the likelihood of biofilm development. (4)

Age and Deterioration:

Aging infrastructure significantly contributes to the biofouling problem. As pipes age, they often corrode, and their surfaces become rougher, making them more susceptible to biofouling. Corrosion can also lead to leaks, which not only waste water but can create environments conducive to biofilm growth both inside and outside the pipe.

The challenge of updating and maintaining aging water systems is considerable. Many urban areas have water systems that are decades old, and the cost of complete replacement is often prohibitively high. Partial upgrades or piecemeal replacements can sometimes inadvertently exacerbate the problem, as new sections of pipe might have different flow characteristics or chemical reactions, creating new areas prone to biofouling.

Regular maintenance and inspection are crucial for mitigating the effects of aging infrastructure. Cleaning and treating pipes to remove biofilms and prevent their reformation is a key part of this maintenance. However, these processes can be disruptive and costly, and in some cases, they may only provide a temporary solution.

In summary, the materials and design of pipes, along with their age and state of deterioration, are central to the challenge of biofouling in water systems. Addressing these issues requires a combination of using improved materials and designs in new installations, regular maintenance and treatment of existing systems, and strategic planning and investment for the upgrade and replacement of aging infrastructure. (5)

Global Examples and Case Studies of Biofouling:

Biofouling is a global issue that affects water systems in various parts of the world, each presenting unique challenges and insights. By examining specific case studies, we can understand the extent of biofouling's impact and the effectiveness of different responses and solutions.

Case Studies

The Thames Water System, London, UK:

In London's Thames Water system, a significant biofouling issue was identified. The city's aging pipe network had become heavily biofouled, leading to water quality concerns and reduced flow efficiency.

Response and Solutions: Thames Water undertook a massive pipe rehabilitation project, relining old pipes with epoxy resin and implementing a regular maintenance program for biofilm removal. This project significantly improved water quality and system efficiency. (6)

The Great Lakes, Canada and the US:

Biofouling in the Great Lakes, primarily due to invasive zebra and quagga mussels, led to clogged pipes and equipment in water treatment plants and distribution systems.

Response and Solutions: The response involved

mechanical and chemical strategies, including the mechanical removal of mussels and chemical treatments like chlorination, as well as redesigning water intake structures. (7)

Desalination Plants in the Middle East:

Desalination plants in the Middle East faced severe biofouling problems due to the warm waters of the Persian Gulf, reducing the efficiency of the desalination process.

Response and Solutions: The plants adopted a combination of pre-treatment processes, including filtration and ultraviolet (UV) light treatment, and applied anti-fouling coatings to surfaces that come into contact with seawater. (8)

Strategies for Mitigation and Prevention of Biofouling

Innovative Technologies:

Advances in nanotechnology have led to surface coatings that resist biofilm formation. Ultrasonic and electromagnetic devices, as well as robotic and automated cleaning systems, offer efficient maintenance solutions. UV-C light technology has shown effectiveness in controlling biofouling by damaging the DNA of microorganisms within biofilms. (9)

Policy and Management Approaches:

Implementing regulatory measures for proactive biofouling management, establishing maintenance protocols, and engaging the community through education about biofouling impacts are essential strategies. These approaches ensure compliance with standards, promote community-driven initiatives for biofouling prevention, and address the critical need of biofouling in our global water systems.

By integrating innovative technologies such as UV-C light, enforcing effective policies, and engaging in proactive management, significant progress can be made toward clean, safe, and efficient water systems. The path forward in managing biofouling lies in innovation, collaboration, and a commitment to sustainable practices, ensuring a healthier future for our water resources.

Footnotes:

1. Biofouling Types and Impacts: "Understanding Biofouling," Water Treatment Journal, 2021.

2. Biofouling and System Efficiency: "Biofouling in Water Systems," Environmental Protection Agency, 2020.

3. Biofilm Resistance to Treatment: "Challenges of Biofilm Control," Water Science and Technology, 2019.

4. Pipe Material Innovations: "Advances in Anti-Biofouling Pipe Coatings," Industrial Coatings World, 2022.

5. Aging Infrastructure and Biofouling: "The Aging Water Infrastructure," Water Systems Council, 2021.

6. Thames Water System Rehabilitation: "Thames Water Pipe Rehabilitation Project," Water and Waste Management, 2004.

7. Great Lakes Biofouling Control: "Invasive Mussel Control in the Great Lakes," Environmental Monitoring and Assessment, 2018.

8. Biofouling in Middle Eastern Desalination Plants: "Biofouling in Desalination Plants," Desalination Research Journal, 2020.

9. UVC Light in Biofouling Prevention: "UVC Technology for biofouling prevention and disinfection"

CHAPTER 7: THE GLOBAL BURDEN OF WATERBORNE DISEASES

The impact of waterborne diseases on global health and economics is profound and far-reaching. According to the World Health Organization, over 3.4 million people die each year as result of water-related diseases; it is the leading cause of disease and death around the world. Most of these victims are young children, with a significant number dying from illnesses caused by organisms thriving in water sources contaminated by raw sewage.

In terms of specific diseases, cholera, typhoid fever, and hepatitis A, caused by bacteria, are among the most common diarrheal diseases. Other illnesses, like dysentery, are caused by parasites found in water contaminated by feces. Annually, there are about four billion cases of diarrhea disease, resulting in approximately one to two million deaths, 90 percent of which tragically occur in children under five years of age. The impact of these diseases is not just limited to health but extends to economic costs. The United Nations has estimated that for every dollar spent on water and sanitation, there is an economic return of between three and thirty-four dollars, depending on the country, due to improved health and productivity.

These numbers underscore the urgency of addressing the water crisis globally, not just as a health issue but also as an economic and developmental challenge. Investments in water treatment, sanitation, and public health measures can significantly reduce the incidence of waterborne diseases and their associated economic burdens. Additionally, education about safe water practices and hygiene can play a crucial role in prevention.

Recognized Waterborne Diseases

Cholera: A severe diarrheal disease caused by Vibrio cholerae bacteria, resulting in dehydration and electrolyte imbalance. Can be fatal if untreated.

Typhoid Fever: caused by Salmonella Typhi bacteria, characterized by fever, headache, constipation, or diarrhea.

Hepatitis A: a viral infection affecting the liver, transmitted through contaminated food or water.

Cryptosporidiosis: caused by Cryptosporidium parasites, leading to respiratory and gastrointestinal illness.

Giardiasis: caused by Giardia parasites, resulting in diarrhea, gas, and stomach cramps.

Norovirus: leading to acute gastroenteritis, characterized by diarrhea, vomiting, and stomach pain.

Shigella: bacteria causing Shigellosis, with symptoms like diarrhea, fever, and stomach pain.

E. coli: certain strains like E. coli O157:H7 cause severe diarrhea and abdominal cramps.

Legionnaires' Disease: A severe form of pneumonia caused by Legionella bacteria.

Leptospirosis: caused by Leptospira bacteria, leading to a wide range of symptoms, sometimes causing severe infections.

Acute Gastroenteritis: characterized by diarrhea, vomiting, and stomach pain.

The global health impact of waterborne diseases is staggering, with a significant portion of the world's population at risk. WHO reports that almost one in three people globally do not have access to safe drinking water, and about 2.4 billion lack access to basic sanitation services. This inadequate access significantly increases the risk of waterborne diseases, which disproportionately affects children and marginalized communities in both urban and rural settings. (1)

Economic impact: Extends far beyond the immediate healthcare costs and loss of human life. It deeply affects the socio-economic fabric of communities, especially in developing countries.

Loss of Productivity: The burden of waterborne diseases significantly impacts economic productivity. Sick individuals cannot contribute to the workforce, leading to a decrease in labour availability and efficiency. This is particularly detrimental in agriculture-based economies where manual labour is critical.

Healthcare Costs: Treating waterborne diseases imposes substantial costs on healthcare systems. In countries where healthcare infrastructure is already under strain, this can lead to overwhelming situations, further straining limited resources.

Impact on Education: Children, often the most affected by waterborne diseases, miss schooling due to illness or because they are tasked with collecting water. This results in a long-term impact on their educational attainment and future economic opportunities.

Gender Disparity: The task of water collection disproportionately falls on women and girls, limiting their time for education, work, or participation in community activities. This gender disparity perpetuates cycles of poverty and limits women's opportunities for economic advancement.

Economic Losses: As per the United Nations Development Programme (UNDP), regions like Africa suffer significant economic losses due to inadequate water supply and sanitation, estimated to be about 5 percent of GDP. This loss is a combination of healthcare costs, lost productivity, and other indirect costs.

Investment in Water Infrastructure: Addressing waterborne diseases requires significant investment in water and sanitation infrastructure. While this is a substantial financial undertaking, the long-term economic benefits of such investments often outweigh the initial costs.

Impact on Tourism and Trade: Countries facing severe waterborne disease challenges may experience reduced tourism and trade. Concerns about health and safety can deter tourists and affect the reputation of local products on the global market.

Long-Term Economic Growth: The persistent challenge of waterborne diseases can hamper long-term economic growth and development. Countries spending a significant portion of their GDP on dealing with these issues may find it challenging to invest in other areas critical for economic development.

The economic impact of waterborne diseases is multifaceted, affecting individual productivity, national healthcare systems, educational outcomes, gender equality, and overall economic growth. Addressing these challenges requires a multi-pronged approach, focusing on improving water and sanitation infrastructure, healthcare facilities, and educational opportunities, especially in the most affected regions.

The lack of clean water and sanitation facilities in many parts of the world not only causes illness but also hinders economic productivity. The time spent collecting water, often by women and children, and the time lost due to illness, detracts from education, work, and care responsibilities. The United Nations Development Programme estimates that the economic losses in Africa due to inadequate water supply and sanitation are about 5 percent of GDP. (2)

Prevention and Control Efforts: Efforts to prevent and control waterborne diseases involve a combination of strategies. These include improving water quality through filtration and disinfection, protecting water sources from contamination, and promoting safe hygiene practices. Governments, international organizations, and NGOs are working together to invest in infrastructure that provides safe drinking water and adequate sanitation.

Prevention and control efforts for waterborne diseases are multifaceted and involve collaboration between governments, international organizations, non-governmental organizations (NGOs), and communities. These strategies are essential for reducing the incidence of these diseases and improving public health.

Improving Water Quality: The primary step is to ensure the provision of safe drinking water. This involves the treatment of water to remove contaminants and pathogens. Methods include filtration, chlorination, and the use of UV radiation. Ensuring access to clean water reduces the reliance on potentially contaminated sources.

Protecting Water Sources: It's crucial to prevent the contamination of water sources. This involves managing waste effectively, implementing agricultural practices that reduce runoff into water bodies, and protecting catchment areas from industrial pollutants.

Safe Hygiene Practices: Educating communities about the importance of handwashing, safe food preparation, and general hygiene plays a significant role in controlling waterborne diseases. Such practices are especially crucial in areas with limited access to clean water.

Infrastructure Development: Investing in infrastructure is vital for long-term solutions. This includes building well-designed water supply systems, adequate sanitation facilities, and effective waste management systems.

Community Engagement and Education: Engaging with local communities to raise awareness about the causes and prevention of waterborne diseases is essential. Community participation ensures the sustainability of initiatives and helps in adapting strategies to local needs.

Surveillance and Reporting Systems: Establishing robust systems for the surveillance of waterborne diseases helps in early detection and response. This includes monitoring water quality and reporting disease outbreaks promptly.

International Collaboration: collaboration at the international level is key to mobilizing resources, sharing best practices, and coordinating efforts, especially in resource-limited settings.

Research and Development: ongoing research into new technologies and methods for water purification, disease detection, and vaccine development is essential for tackling emerging challenges in waterborne disease control.

These efforts are vital for reducing the burden of waterborne diseases globally, especially in regions where access to clean water and sanitation facilities is limited. Effective prevention and control of these diseases contribute significantly to improving public health and economic productivity.

Global Initiatives: Several global initiatives aim to address the issue of waterborne diseases. The United Nations' Sustainable Development Goals (SDGs), specifically Goal 6, focuses on ensuring the availability and sustainable management of water and sanitation for all. WHO and UNICEF's Joint Monitoring Programme (JMP) works toward monitoring progress on drinking water, sanitation, and hygiene.

Challenges in Implementation: Despite these efforts, there are significant challenges in the implementation of water and sanitation programs. These include limited financial resources, weak institutional capacity, and lack of public awareness. Addressing these challenges requires a multi-sectoral approach involving governments, civil society, and the private sector.

The burden of waterborne diseases is a critical global health issue that requires immediate and sustained attention. The intersection of health, economics, and human rights in this issue underscores the need for comprehensive strategies that address the root causes of water and sanitation-related diseases. Collaborative efforts across various sectors are vital to reduce the incidence of these diseases and improve the quality of life for millions of people worldwide. (3)

Footnotes:

1. WHO and UNICEF: "Progress on drinking water, sanitation and hygiene: 2017 update and SDG baselines."

2. United Nations Development Programme: "Human Development Report 2006 - Beyond scarcity: Power, poverty and the global water crisis."

3. United Nations: "Sustainable Development Goal 6 Synthesis Report 2018 on Water and Sanitation."

CHAPTER 8: THE RISING TIDE: CLIMATE CHANGE AND WATERBORNE DISEASES

A changing climate altered weather patterns, extreme events, and temperature fluctuations; these changes are reshaping our ecosystems, not as statistical anomalies, but as pivotal shifts in our environment.

Focusing on water, the lifeblood of our communities, we see how climate change impacts water resources multifacetedly. Rivers at times swell with floodwaters, carrying pollutants and pathogens, and at other times reduce to trickles, concentrating harmful substances. This dichotomy threatens water quality and availability, crucial for human survival.

Our understanding of the relationship between climate change and waterborne diseases is underpinned by robust scientific research. Numerous studies have elucidated a clear link between rising global temperatures and the increase in incidences of various waterborne diseases. (1)

One notable example is the research on cholera. Studies have shown that both the prevalence and spread of cholera are influenced by changes in rainfall patterns and temperature. Warmer temperatures accelerate the growth of Vibrio cholerae, the bacterium responsible for cholera, particularly in water bodies. Furthermore, heavy rainfall and flooding can disrupt sanitation systems, leading to the contamination of drinking water with cholera bacteria. (2)

Similarly, research on E. coli infections has demonstrated that during periods of drought, reduced water flows in rivers and streams lead to higher concentrations of these bacteria. The decrease in water volume concentrates the pollutants, including pathogenic bacteria, making water sources more hazardous. (3)

The Role of Ocean Warming

Ocean warming, a consequence of climate change, also significantly impacts the proliferation of waterborne diseases. Warmer sea temperatures have been linked to the increased presence of pathogens like Vibrio species, which can cause illnesses ranging from gastroenteritis to severe systemic infections. The warming oceans expand the habitable zone for these pathogens, leading to a broader geographical spread of diseases. (4)

The Impact of Extreme Weather Events

Extreme weather events, which are becoming more frequent and severe due to climate change, also play a critical role in the spread of waterborne diseases. Heavy rainfall and flooding, for instance, often result in the contamination of water sources with runoff containing pathogens. On the other hand, extreme heat and drought conditions can reduce water availability, leading to the use of unsafe water sources, thereby increasing the risk of disease transmission.

We recount instances like the cholera outbreaks following major floods in South Asia. These outbreaks vividly illustrate how the combination of water contamination and disrupted infrastructure can lead to widespread disease. As noted earlier, the case of Walkerton, Ontario, where heavy rainfall led to E. coli contamination of the town's water supply, resulting in a tragic outbreak, is another poignant example.

In regions experiencing droughts, such as certain parts of Africa and Australia, there has been an observed increase in waterborne diseases due to the use of contaminated or untreated water sources. These examples highlight the direct impact of climate change on public health and the importance of resilient water management and healthcare systems. Our understanding of the relationship between climate change and waterborne diseases is underpinned by robust scientific research.

Numerous studies have elucidated a clear link between rising global temperatures and the increase in incidences of various waterborne diseases. One notable example is the research on cholera. Studies have shown that both the prevalence and spread of cholera are influenced by changes in rainfall patterns and temperature. Warmer temperatures accelerate the growth of Vibrio cholerae, the bacterium responsible for cholera, particularly in water bodies. Furthermore, heavy rainfall and flooding can disrupt sanitation systems, leading to the contamination of drinking water with cholera bacteria.

The Rising Threat: Legionella and Other Waterborne Pathogens

Climate changes foster environments conducive to pathogens like Legionella; Legionnaires' disease thrives in warm, stagnant water found in both natural environments and in man-made water systems, increasingly prevalent as global temperatures rise. Urban heat island effects exacerbate this risk, with epidemics documented in cities experiencing heatwaves. Instances of Legionnaires' disease outbreaks have also been linked to climate-induced environmental changes.

Other pathogens like Vibrio species, Cryptosporidium, Giardia, and harmful algal blooms are on the rise as well, also influenced by climate change. Warmer sea temperatures expand Vibrio habitats, while heavy rainfall and flooding increase the prevalence of Cryptosporidium and Giardia. Warming waters lead to harmful algal blooms, producing toxins affecting human health.

Vibrio Species: Thriving in warmer waters, Vibrio species, particularly known for causing cholera, have been a focal point of research in understanding climate change's impact on waterborne pathogens. These bacteria naturally inhabit coastal waters, and their growth is highly temperature dependent. As sea temperatures rise due to global warming, Vibrio species find more hospitable conditions, expanding both in population size and geographical range.

The implications of this expansion are significant. For instance, regions previously considered at low risk for Vibrio-related diseases are now experiencing outbreaks. Studies have shown an increased incidence of Vibrio infections in areas where sea temperatures have risen, even by small margins. The increase in Vibrio pathogens not only affects human health through diseases like cholera and gastroenteritis but also impacts marine life and ecosystem health.

Cryptosporidium and Giardia: These are protozoan parasites causing gastrointestinal illnesses, exhibiting a different pattern of climate-related proliferation. Both parasites are commonly transmitted through water contaminated with feces and are notorious for causing outbreaks in communities with inadequate water treatment facilities.

Climate change contributes to the spread of these parasites primarily through increased rainfall and flooding. Heavy rains can overwhelm sewage systems and agricultural runoffs, leading to the contamination of water bodies, including sources of drinking water. Flooding events can spread these parasites over large areas, making containment and treatment challenging. Cryptosporidium, in particular, is highly resistant to chlorine, which is commonly used in water treatment. This resistance, combined with increased flooding events, poses a substantial risk to drinking water safety, especially in areas without advanced water treatment capabilities.

Algal Blooms: Algal blooms, particularly harmful algal blooms (HABs), have garnered attention as a significant environmental concern exacerbated by climate change and increased nutrient runoff from agriculture. These blooms occur when colonies of algae—simple plants living in the sea and freshwater—grow out of control and produce toxic effects on people, fish, shellfish, marine mammals, and birds. The frequency and severity of HABs are increasing with warming water temperatures, as algal blooms thrive in warm, nutrient-rich environments.

Climate change contributes to this by not only increasing water temperatures but also by altering patterns of rainfall and runoff, thereby increasing the flow of nutrients into water bodies. The toxins produced by some algal blooms can have severe health effects on humans, ranging from respiratory problems, liver damage, and serious neurological effects when consumed. Furthermore, they can decimate local fish populations, threaten water safety, and disrupt aquatic ecosystems, leading to broader ecological consequences.

Mitigating the Impact: Addressing the growing threat posed by these pathogens requires a multifaceted approach. Monitoring and early detection of these pathogens are critical, as is investment in water treatment infrastructure to manage and mitigate their spread. Public health initiatives focusing on education about the risks associated with contaminated water, particularly in vulnerable regions, are also essential.

Moreover, global efforts to mitigate climate change can indirectly help in controlling the proliferation of these waterborne pathogens. By curbing the temperature increases and extreme weather events that facilitate the spread of these pathogens, we can reduce the incidence of the diseases they cause.

Spotlighting Vibrio species, Cryptosporidium, Giardia, and harmful algal blooms in the context of climate change highlights the intricate and often overlooked ways in which environmental changes impact human health. Understanding these relationships is crucial for developing effective public health strategies and policies to safeguard communities against the rising tide of waterborne diseases in a changing climate.

The connection between extreme weather events, such as heavy rainfall, floods, and heatwaves, and the proliferation of waterborne diseases, precipitation patterns, and temperature fluctuations, additionally contribute to a higher risk of waterborne diseases.

I believe that the significant health implications of these climate-induced changes in waterborne disease patterns is highlighting the strain on healthcare systems, especially in regions with infrastructural challenges.

Forward-Looking Strategies

Enhanced water treatment facilities, rigorous monitoring and maintenance of water systems, public education about water safety, and investments in research and early warning systems are needed.

Public Health Initiatives and Technology

The role of public health initiatives in mitigating the impact of waterborne diseases and the potential of technological advancements in managing these risks need to be explored.

Footnotes:

(1): Hunter, Paul R. "Climate change and waterborne and vector-borne disease." Journal of Applied Microbiology

(2): Lipp, Erin K., Anwar Huq, and Rita R. Colwell. "Effects of global climate on infectious disease: the cholera model." Clinical Microbiology Reviews

(3): McLellan, Sandra L., et al. "Distribution and fate of Escherichia coli in Lake Michigan following contamination with urban stormwater and combined sewer overflows." Journal of Great Lakes Research

(4): Vezzulli, Luigi, et al. "Ocean warming and spread of pathogenic vibrios in the aquatic environment."

CHAPTER 9: RISK MANAGEMENT IN REAL ESTATE: NAVIGATING THE WATERS OF LIABILITY IN WATERBORNE ILLNESSES

The real estate industry, while focused on investments and development, often underestimates a critical hazard: the risk associated with waterborne illnesses, especially Legionnaires' disease. This chapter emphasizes the financial, legal, and reputational repercussions for real estate entities neglecting this risk, focusing on Legionnaires' disease across the United States. (1)

Legionnaires' disease, caused by Legionella bacteria, typically proliferates in large building water systems, posing a significant risk to hotels and apartment complexes. This disease transcends health hazards, embodying a legal and financial threat to property owners and managers. (2)

The New York Times

Latest Legionnaires' Outbreak in the Bronx Kills 1 and Sickens 18

Health officials urged Bronx residents with flulike symptoms to seek medical care after Legionella bacteria were found in cooling towers there.

🎁 Share full article ↝ ◻

By **Sharon Otterman**

May 25, 2022

A growing cluster of Legionnaires' disease cases in the Bronx has resulted in the death of one person and illness in 18 others, eight of whom are currently hospitalized, the New York City Health Department said Wednesday.

The cases appear to be linked to four water cooling towers on top of buildings in the Highbridge neighborhood of the Bronx, where officials said they found Legionella pneumophila, the bacteria that causes Legionnaires' disease.

Outbreaks and Consequences:

The Gotham Hotel Outbreak, New York City:

In 2017, two guests contracted Legionnaires' disease, one fatally, at this upscale three-hundred-room hotel in Manhattan.

Financial Outcome: over $5 million in lawsuits, reputation damage, and booking declines. (3)

Riverside Apartments Incident, Chicago:

A 2018 outbreak at this 250-unit apartment complex affected multiple elderly tenants.

Financial and Legal Consequences: lawsuits exceeding $10 million; significant remediation costs. (4)

Sunshine Towers Hotel Case, Las Vegas:

There were twenty confirmed cases in 2019 at this four-hundred-room hotel catering to tourists and business travellers.

Financial Impact: lawsuits over $15 million; revenue loss due to temporary closure. (5)

Bayview Condominiums Health Scare, Miami:

Several residents contracted the disease in 2020 at this two-hundred-unit luxury condominium complex.

Costs Incurred: legal claims over $8 million; extensive renovations and upgrades.

Downtown Residences Outbreak, Seattle:

A 2021 outbreak affected many residents at a 350-unit luxury apartment building.

Financial Repercussions: around $12 million in lawsuits; system rectification costs and value impact.

Reputation Risk: The Invisible Threat

In the realm of real estate, the reputation of a property or company is as valuable as its physical assets. An outbreak of a waterborne illness like Legionnaires' disease can lead to immediate and long-lasting damage to a property's reputation. The news of an outbreak can spread rapidly, causing public fear and mistrust. This negative publicity can lead to:

Decreased Occupancy: potential residents or guests may choose alternative accommodations, leading to lower occupancy rates and revenue loss.

Reduced Property Value: properties known for health risks can see a decline in market value and attractiveness to potential buyers or investors.

Long-term Brand Damage: for real estate companies, an outbreak can tarnish the entire portfolio's reputation, impacting future projects and investor confidence.

Legal and Financial Costs: besides the direct costs of litigation, the indirect costs of managing a tarnished reputation can be substantial, including PR campaigns, increased marketing expenses, and potential concessions to retain clientele.

Safeguarding Against Risks: The cases highlighted in

this chapter underscore the multifaceted impact of waterborne illness outbreaks. Real estate executives must recognize and act upon the critical importance of managing these risks to protect their tenants, properties, and companies from the far-reaching consequences of such incidents.

Implementing rigorous water safety protocols and maintaining up-to-date infrastructure are not just regulatory compliances but essential strategies for safeguarding against financial, legal, and reputational risks. The cost of implementing preventive measures is far outweighed by the potential costs of inaction.

Footnotes:

(1): Garrison, Laura E., et al. "Legionnaires' disease incidence and risk factors, New York, New York, USA, 2002–2011." Emerging Infectious Diseases

(2): Hamilton, Kerry A., et al. "Risk-based critical control point identification and the significance of non-culturable Legionella pneumophila in premise plumbing." Environmental Science: Water Research & Technology

(3): Fitzhenry, Robert, et al. "Legionnaires' Disease Outbreaks and Cooling Towers, New York City, New York, USA." Emerging Infectious Diseases

(4): Jones, Timothy F., et al. "Legionnaires' disease outbreak investigation team: An outbreak of Legionnaires' disease associated with a decorative water wall fountain in a hospital." Infection Control & Hospital Epidemiology

(5): Beer, Karlyn D., et al. "Outbreaks associated with environmental and undetermined water exposures — United States, 2011–2012." Morbidity and Mortality Weekly Report

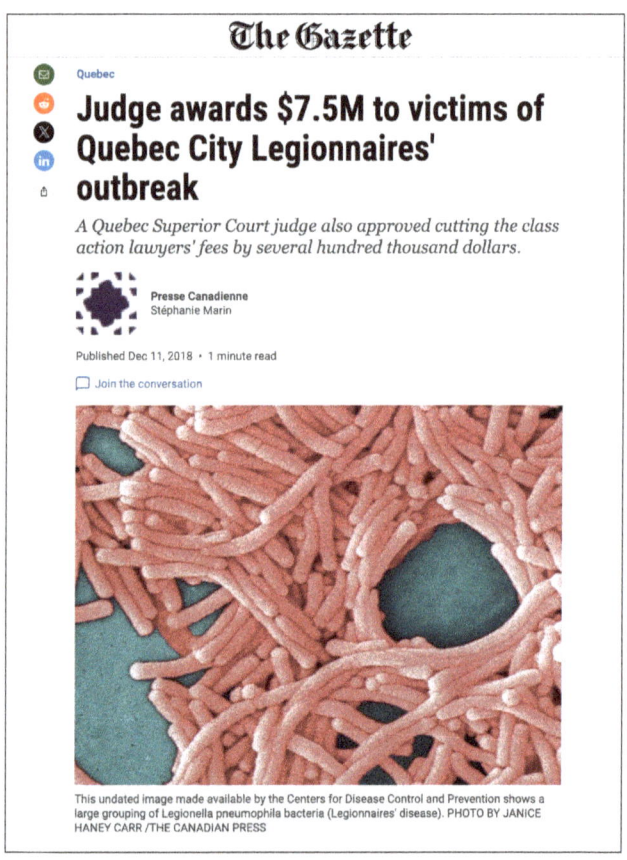

The Gazette

Quebec

Judge awards $7.5M to victims of Quebec City Legionnaires' outbreak

A Quebec Superior Court judge also approved cutting the class action lawyers' fees by several hundred thousand dollars.

Presse Canadienne
Stéphanie Marin

Published Dec 11, 2018 · 1 minute read

Join the conversation

This undated image made available by the Centers for Disease Control and Prevention shows a large grouping of Legionella pneumophila bacteria (Legionnaires' disease). PHOTO BY JANICE HANEY CARR /THE CANADIAN PRESS

CHAPTER 10: ELEVATING REAL ESTATE THROUGH ESG-DRIVEN WATER MANAGEMENT INITIATIVES

Environmental, Social, and Governance (ESG) principles are increasingly becoming pivotal in shaping real estate strategies. This chapter provides an in-depth exploration of how ESG-centric water management, particularly through water sterilization systems, is crucial in real estate development and management, addressing environmental sustainability, social responsibility, and governance adherence. (1)

The Environmental Aspect: Sustainable Water Practices

The environmental dimension of ESG in real estate revolves around sustainable water practices.

Integrating water sterilization systems is a key component of this approach, contributing to environmental conservation in several ways:

Reduction of Chemical Usage: water sterilization systems significantly reduce reliance on harmful chemicals, thereby minimizing ecological damage and preserving biodiversity in local water bodies.

Water Conservation and Recycling: implementing advanced water treatment technologies, including sterilization and recycling systems, leads to efficient water use and conservation, a crucial factor in sustainable building practices.

Energy Efficiency and Reduced Carbon Footprint: optimizing water management through sterilization systems can decrease energy consumption associated with water heating and treatment, contributing to the reduction of a building's overall carbon footprint.

The Social Dimension: Health, Safety, and Community Engagement Social responsibility in water management focuses on the well-being of tenants and the community:

Health and Safety Assurance: by implementing water sterilization systems, real estate companies can ensure the provision of safe and clean water, thereby safeguarding tenant health and preventing waterborne diseases.

Educational Initiatives and Community Involvement: engaging with tenants and the local community through educational programs on water conservation and safety fosters a sense of responsibility and collective action toward environmental stewardship.

Enhanced Living Conditions: providing high-quality water contributes to better living conditions, directly impacting the quality of life and tenant satisfaction.

Governance: Ethical Management and Regulatory Alignment Governance in ESG-focused water management encompasses compliance with laws and ethical practices.

Adherence to Water Quality Regulations: ensuring that water management systems meet local and international regulatory standards is fundamental for legal compliance and risk mitigation.

Ethical Management and Transparency: ethical management practices, including transparent reporting on water management policies and practices, are essential to build trust with investors, tenants, and regulatory bodies.

Proactive Risk Management: implementing water sterilization systems as a preventive measure against potential waterborne hazards demonstrates responsible and foresighted governance.

I believe that the future of water management in real estate under ESG principles will likely witness several evolving trends:

Adoption of Cutting-Edge Technologies: innovations such as AI-driven water management systems and advanced sterilization technologies will become more prevalent, offering enhanced efficiency and monitoring capabilities.

Cross-Sector Collaborations: partnerships between real estate companies, technology firms, and environmental organizations will drive the development of sustainable water management solutions.

Global Benchmarking and Standardization: as ESG becomes a global imperative, standardized benchmarks for sustainable water management in real estate are expected to emerge, guiding industry practices worldwide.

Forging a Sustainable Path in Real Estate

The integration of ESG principles, particularly in water management, is not a mere trend but a necessity in the contemporary real estate landscape. By adopting sustainable water practices, real estate entities not only contribute to environmental conservation and social well-being but also adhere to ethical governance standards. This chapter has underscored the importance of proactive ESG-driven water management strategies, encouraging real estate leaders to embrace these practices for long-term sustainability, profitability, and societal impact. The future of real estate is inextricably linked to sustainable practices, and water management is at the heart of this evolution.

Footnotes:

(1): Eccles, Robert G., and Svetlana Klimenko. "The Investor Revolution: Shareholders are getting serious about sustainability." Harvard Business Review, May–June 2019.

(2): Maas, Carolin, et al. "Environmental Management and Sustainability in Real Estate: A Review of Current Practices and Future Outlook." Journal of Property Investment & Finance

(3): Global Real Estate Sustainability Benchmark (GRESB). "2021 GRESB Real Estate Assessment Reference Guide."

CHAPTER 11: CLEAR INC.: TRANSFORMING GLOBAL WATER SAFETY

My journey with Clear Inc. began not merely as a quest for technological advancement but as a moral imperative to ensure safe indoor environment for everyone.

Thirteen years ago, I started to research for ways to do environmental indoor building checks; the idea was to visit homes once a year, taking air water pills and other environmental indicators, bring them back to analyze, give the homeowner a certificate proving the house was safe and clear, and if there were any environmental health concerns, direct the homeowner to the right professionals that could deal with the challenge.

I spent countless hours developing the idea, but I

understood that there was one major caveat—indoor environmental indicators change on a daily basis, so the fact that we did the check on Sunday and then provided the information to the homeowner on Wednesday, the environment could have completely changed in that time; for example, a crack in the neighbourhood watermain pipe. I decided to freeze the concept until more technological advancements were introduced that would allow us to monitor indoor environments twenty-four-seven, not only once a year.

In 2019, my son Ron's illness changed it all for me. Ron contracted a rare but lethal bacterium from a glass of water during a trip in Europe. This bacterium, responsible for over 130,000 global deaths annually and chronic bowel problems in countless survivors, changed his life and mine. Witnessing Ron's transformation from a vibrant, healthy individual into someone grappling with constant discomfort and anxiety was heart-wrenching. His struggle with persistent diarrhea and stomach pains lasted three years before his health stabilized.

Ron, a brilliant PhD student in genetics, had dedicated his research to extending human lifespan through DNA manipulation. However, his illness brought a shift in his perspective. He began to see that longevity without quality of life was meaningless. It was during this challenging period that Ron expressed his desire to join me in finding a large-scale solution to waterborne diseases.

We started with comprehensive research on waterborne diseases in Canada, based on data from the Canadian health ministry. The severity of biological contamination in our drinking water was far worse than we had imagined. (1)

Ron and I understood the biofouling problem in the distribution system; the pipes that go from the water treatment plants to our homes is the main cause of most of these diseases, and biofouling is a natural phenomenon that continues to grow viruses, bacteria and pathogens, and it is almost impossible to deal with.

We learned that to replace the city pipes was almost mission impossible due to the cost, and we understood that the only solution was to install a firewall between the city pipes that would penetrate our buildings to the rest of the pipes in the building systems and prevent any biological contamination from being introduced from the city pipes.

Together, we began two years of research for the best systems that were already proven and known in the market. We decided to follow the global-leading pharmaceutical and food and beverage companies and examined the machines needed to transfer the city water into microbiological safe water in the factories.

One name came coming up, Atlantium Technologies, an Israeli company that for the last twenty years had been servicing most of the top pharmaceutical and food and beverage companies worldwide. (2) Doing the technical due diligence on their solution, we uncovered a story of an Israeli father and son that were committed to making the world a better and safer place.

The Visionaries: In the world of water treatment and safety, Benjamin Kahn and Morris Kahn of Atlantium Technologies stand as beacons of innovation and environmental stewardship. Benjamin, influenced by his father Morris, a renowned entrepreneur and philanthropist, embarked on a journey to revolutionize water treatment through Atlantium. His vision was not just about technological breakthroughs; it was deeply intertwined with environmental consciousness.

Beyond Atlantium, both Benjamin and Morris Kahn have been influential figures in other significant ventures. Notably, Morris was involved in the founding of Amdocs, a leading software and services provider for communications and media companies. Furthermore, in 2007, Benjamin was recognized by *Time* magazine as a "Hero of the Environment," highlighting his contributions to environmental conservation.

One of their most ambitious projects was the Genesis lunar mission, an endeavour to land an Israeli spacecraft on the moon. This mission, though not fully successful in its lunar landing, was a testament to their dedication to pushing the boundaries of technology and exploration.

Morris Kahn contributed significantly to their technological advancements with his expertise in environmental sciences and a deep understanding of waterborne diseases. This expertise was pivotal in developing Atlantium's unique Hydro-Optic™ (HOD) UV technology Atlantium's HOD UV system was a leap forward in ultraviolet water treatment, combining hydraulics and fibre-optic principles for efficient delivery of UV light. This system ensured effective neutralization of waterborne pathogens, offering a higher level of biosecurity. Its efficiency and cost-effectiveness made it a game changer, minimizing energy consumption and operational costs while providing an eco-friendly solution.

Under the leadership of Benjamin and Morris, Atlantium's technology was adopted in over four thousand installations across sixty-three countries. Their commitment to sustainability further enhanced Atlantium's global appeal, providing a non-chemical, environmentally friendly approach to water treatment.

After my visit to the Atlantium Technologies factory in Beit Shemesh, Israel, I knew that I had found what I was looking for—a mature company that was trusted by main industrial companies worldwide. It seemed that every employee in the factory and in the offices was fully committed to producing the best technology and making the world a healthier place.

Atlantium's management team, led by the CEO Yariv Abramovitz and Senior VP Shai Goren, understood our mission and for a few months, we negotiated a global exclusivity deal to install Atlantium's biological firewall in residential and commercial communities across the globe.

The early stages of Clear Inc.'s development, while promising, do not yet fully reveal the company's potential for success. The need for our services is evident, and our approach is both innovative and lucrative. Our initial installations in cities like Montreal, Toronto, Miami, Manila, and across California, Washington, and Fort Lauderdale are early indicators of the demand for our technology. These installations reflect a growing awareness of environmental challenges and health concerns, positioning Clear at the forefront of a significant shift in water safety regulations.

I firmly believe that in the future, regulations will likely mandate systems like ours in buildings for water disinfection. This belief stems from the increasing concerns about global warming, water scarcity, and the environmental and health issues discussed throughout this book. The trajectory toward mandatory water safety measures seems clear, driven by a global movement toward environmental sustainability and public health.

Powered by Atlantium, Clear transcends being a mere water treatment system; it symbolizes a future where access to safe water is a universal right. Our mission is global installation, starting with early adopters who are integral to this transformative journey.

My son's illness and recovery, combined with our encounter with Atlantium's visionary technology, have reshaped our approach to water safety. This chapter is not just the story of a business venture; it's a narrative of a father and son, a family changed by adversity, and their contribution to a global movement for a healthier future.

Footnotes:

(1): Hrudey, Steve E., and Elizabeth J. Hrudey. "Safe Drinking Water: Lessons from Recent Outbreaks in Affluent Nations."

(2): Atlantium Technologies Ltd. "About Atlantium." Atlantium Technologies Ltd, 2021.

CHAPTER 12: MAXIMIZING PROFIT AND SUSTAINABILITY THROUGH CENTRAL WATER PURIFICATION SYSTEMS IN REAL ESTATE

As a property owner and Investor that is committed to sustainability and health, I have seen firsthand how the real estate landscape is rapidly evolving. Today's buyers and renters are increasingly looking for properties that offer not only comfort and convenience but also a commitment to health and environmental sustainability.

Integrating central water purification systems, especially those that leverage advanced technologies like Ultraviolet-C (UVC), presents a unique opportunity for property owners and developers to meet these demands. This chapter delves into the financial, environmental, and health benefits of these systems in the real estate market, drawing from my personal experiences and insights. The Shift Toward Health-Centric Real Estate In today's real estate market, there's a clear shift towards health-centric living environments. Consumers are more informed and concerned about the quality of their living spaces than ever before. They demand homes that promote well-being, and central water purification systems play a crucial role in meeting this demand. By ensuring a safe and high-quality water supply, these systems enhance the overall appeal of a property. One pivotal moment for me was when I considered installing a central water purification system in one of my properties. It was a significant investment, but the feedback from potential tenants was overwhelmingly positive and I succeeded to lease up the spaces very fast. The new tenants appreciated the commitment to their health and safety, which in turn increased their satisfaction and loyalty.

I believe that Energy Efficiency and Cost Saving is One of the most compelling advantages of central water purification systems is their ability to reduce energy costs. By maintaining lower water temperatures and preventing scale and biofilm formation in pipes, these systems contribute to substantial energy savings and lower operational costs (2). In many cases, the initial investment in a high-quality water purification system quickly paid off through reduced utility bills and maintenance costs.

The improved efficiency of the plumbing infrastructure also meant fewer disruptions for tenants, which is always a win for property management. Maintenance and Operational Efficiency. Central water purification systems significantly reduce the need for frequent maintenance by preventing pipe corrosion and scaling. This leads to longer-lasting plumbing infrastructure and lower repair costs, contributing to overall operational efficiency (3). There was a period when pipe corrosion was a recurring issue in one of my friend's building. After installing a water purification system, the reduction in maintenance issues was noticeable almost immediately. Not only did this save on repair costs, but it also minimized the inconvenience to tenants, enhancing their living experience.

Enhanced Property Value and Market Appeal Properties equipped with central water purification systems, particularly those utilizing UV-C technology, tend to attract a broader demographic and often command higher market values. This reflects the growing consumer preference for health-focused and sustainable living spaces. When I listed a property featuring a state-of-the-art water purification system, the response from Potential tenants was impressive. They saw it as a significant value addition, which set the property apart from others on the market. It was clear that such features are no longer a luxury but a necessity for many health-conscious buyers. Reduction in Bottled Water Usage central water purification systems also significantly reduce the dependency on bottled water, aligning with environmental sustainability goals and reducing operational expenses.

In properties that installed central systems the switch to purified tap water has been economic and environmental boon. Tenants appreciated not having to purchase bottled water, and the reduction in plastic waste was a meaningful contribution to sustainability efforts.

Regulatory Compliance and Risk Mitigation Using central water purification systems helps property owners comply with stringent health and safety regulations, mitigating the risk of legal liabilities associated with waterborne diseases (1).

Navigating the regulatory landscape can be challenging, but having a water purification system in place has provided peace of mind. It ensures compliance with health standards and significantly reduces the risk of tenant complaints or health issues related to water quality.

Environmental Impact and Sustainability: These systems contribute to environmental sustainability by reducing energy consumption and lowering the carbon footprint of properties. For someone deeply committed to sustainability, this aspect is particularly important to me. The positive environmental impact of these systems aligns with my personal and professional values, making them a key feature in all my real estate investments. Consumer Preference and Willingness to Pay There's a growing consumer trend toward properties with health-enhancing features, with many willing to pay a premium for such amenities (4). I've observed that properties with advanced water purification systems not only attract more interest but also justify higher rental rates or sale prices. People are increasingly willing to invest in their health and well-being, and this trend shows no signs of slowing down.

The Future of Real Estate: Integrating Advanced Water Purification

As I look to the future of real estate, I see central water purification systems becoming a standard feature in modern, health-conscious, and environmentally responsible properties. The integration of these systems is not just a trend but a necessary evolution in how we think about property development and management.

Case Study: MIT Center for Real Estate Report on Healthy Buildings

The MIT Center for Real Estate Research Paper

No. 21/04 provides valuable insights into the financial impact of health-centric features in commercial real estate. According to this report, properties that incorporate health-focused features, such as advanced water purification systems, command between 4.4% and 7.7% higher effective rents compared to non-certified peers. This premium is independent of factors like LEED certification, building age, renovation, lease duration, and submarket. MIT's analysis demonstrated that tenants are willing to pay a premium for buildings that offer healthier environments, which include high-quality water and air systems. The report highlighted that these health-focused measures not only enhance tenant satisfaction but also reduce vacancy rates, thereby increasing the overall market appeal and value of the properties. Community and Environmental Benefits Beyond the immediate financial and operational benefits, the installation of these systems also contributed positively to the community and the environment. The reduction in plastic waste from bottled water alone was significant. Additionally, the building's overall environmental footprint decreased, aligning with broader sustainability goals. Personal Commitment to Health and Sustainability My commitment to health and sustainability in real estate is deeply personal. Having faced my son Ron health challenges due water-born bacteria, I understand the importance of a safe and healthy living environment. This drives my passion for integrating advanced water purification systems in all my properties. It's about creating spaces where people can thrive, knowing their health and well-being are prioritized.

I believes that Central water purification systems represent a forward-thinking strategy in real estate, offering financial benefits, enhancing property appeal, and contributing to environmental conservation. As the industry continues to evolve, these systems are likely to become a standard in modern, health-conscious, and environmentally responsible properties. Reflecting on my journey, the decision to integrate these systems into my properties has been transformative. It's not just about staying ahead of market trends; it's about creating spaces that prioritize the health and well-being of residents while also being mindful of our environmental impact. As more property owners recognize these benefits, I believe we will see a significant shift toward more sustainable and health-focused real estate practices.

Footnotes:

(1): World Health Organization. "Guidelines for Drinking-water Quality, 4th Edition." WHO, 2011.

(2): U.S. Environmental Protection Agency. "A Guide to Energy-Efficient Heating and Cooling." EPA, 2009.

(3): International Ultraviolet Association. "Ultraviolet Disinfection Guidelines for Drinking Water and Water Reuse." IUVA, 2003.

(4): National Association of Realtors. "2020 Profile of Home Buyers and Sellers." NAR, 2020.

CHAPTER 13: THE FUTURE OF WATER MANAGEMENT: INNOVATIONS AND CHALLENGES

The realm of water management is on the cusp of a revolution, driven by technological innovation. This chapter embarks on an exploration of emerging trends and the associated challenges in water management, focusing on nanotechnology, artificial intelligence and machine learning, decentralized systems, and water recycling and reuse.

Nanotechnology in Water Purification

Nanotechnology has emerged as a powerful tool in water purification, offering innovative and efficient solutions to address global water challenges. This section delves into the specifics of how nanotechnology is revolutionizing water purification, the key players in the industry, and the challenges and risks associated with these advancements.

The Mechanism and Applications: Nanotechnology in water purification primarily involves the use of nano-filters and nano-catalysts. Nano-filters, with their minuscule pore sizes, can effectively remove even the smallest contaminants, including bacteria, viruses, and heavy metals. (1) These filters operate at a molecular level, providing a higher degree of purification compared to conventional filtration methods. Nano-catalysts, on the other hand, are used to break down pollutants through advanced chemical processes. They are particularly effective in degrading organic compounds that are otherwise difficult to remove. (2)

Leading companies such as NanOasis in the US and NanoSun in the UK are at the forefront of this technology. NanOasis specializes in creating nano-structured membranes that dramatically enhance water throughput while effectively filtering out contaminants. (3) NanoSun, meanwhile, focuses on developing portable water purification devices that utilize nanotechnology, making clean water accessible even in remote locations. (4)

Challenges and Risks: Despite the remarkable benefits, nanotechnology in water purification comes with its own set of challenges. One of the primary concerns is the potential release of nanoparticles into the environment. These particles, due to their size, can easily penetrate biological membranes, potentially leading to unforeseen environmental and health risks. (5) There is ongoing research to understand the long-term impact of nanoparticles on ecosystems and human health.

Scalability is another significant challenge. While nanotechnology offers exceptional benefits in laboratory settings, replicating these results on a larger scale, such as in municipal water treatment plants, poses considerable technical and economic challenges. The high cost of nanomaterials and the complexity of manufacturing nano-filters and nano-catalysts at a large scale are obstacles that need to be addressed.

The future of nanotechnology in water purification looks promising, with continuous research and development paving the way for more sustainable and safe applications. To overcome the current challenges, it is essential to develop standardized regulations for the use of nanotechnology in water treatment and invest in research that assesses the long-term impacts of nanoparticles. Collaborative efforts between government bodies, research institutions, and private companies will be crucial in advancing this technology in a way that maximizes its benefits while minimizing potential risks.

Artificial Intelligence and Machine Learning in Water Quality Monitoring

In the realm of water quality monitoring, Artificial Intelligence (AI) and Machine Learning (ML) have become game changers. These technologies offer advanced capabilities to analyze and interpret vast amounts of data, leading to more accurate and timely assessments of water quality. This expanded exploration will delve into how AI and ML are transforming water quality monitoring, the pioneering companies in this field, and the challenges associated with these technologies.

Advanced Capabilities and Real-World Applications:

AI and ML algorithms can process complex datasets from various sources, such as sensors, satellites, and historical records. These technologies can identify patterns and trends that might be invisible to human analysts. They can predict potential contamination events, assess the effectiveness of water treatment processes, and even guide decision-making in water management.

Leading companies in this sector include BWT (Best Water Technology) Group based in Austria and the SUEZ Group from France. BWT is innovating AI-driven water treatment solutions, enhancing efficiency and sustainability in water management. SUEZ Group harnesses AI and ML for smarter water and waste management, using these technologies to optimize treatment processes and improve water distribution systems.

Challenges and Risks in AI-Driven Water Quality Monitoring: While AI and ML offer incredible potential, they also come with significant challenges:

Data Privacy and Security: The vast amounts of data collected and processed by AI systems raise concerns about privacy and security. Ensuring that sensitive information, particularly in regions with strict data protection laws, is securely handled is paramount.

Accuracy and Reliability of AI Predictions: The reliability of AI-driven predictions is crucial for water quality assessment. Inaccurate predictions can lead to false alarms or, worse, failure to detect real threats. The quality of the data fed into AI models significantly impacts their accuracy. Biased or incomplete data can skew results, leading to potentially misleading conclusions.

Integration with Existing Systems: Integrating AI and ML into existing water monitoring infrastructure can be challenging. It often requires significant investment and expertise to ensure seamless integration without disrupting current operations.

Dependence on Technology: Over-reliance on AI and ML might lead to complacency in manual monitoring and maintenance. Balancing technological and human oversight is necessary to ensure a comprehensive approach to water quality management.

Future Prospects and Directions: The future of AI and ML in water quality monitoring is highly promising. As technology evolves, we can expect more sophisticated models that offer even greater accuracy and efficiency. However, addressing the challenges of data privacy, accuracy, integration, and balancing technology with human oversight is essential for the sustainable adoption of AI and ML in this field.

Continued investment in research and development, coupled with collaboration between technology providers, water industry experts, and regulatory bodies, will be crucial. Developing robust frameworks for data management and model validation, and ensuring the upskilling of personnel in AI and ML applications, will pave the way for more reliable and effective water quality monitoring systems. (6)

This comprehensive exploration underscores the transformative impact of AI and ML in water quality monitoring. By overcoming the current challenges, these technologies can significantly enhance our ability to maintain safe and clean water, a vital resource for all.

Decentralized Water Treatment Systems

The shift toward decentralized water treatment systems marks a significant evolution in the approach to water management. These systems offer localized solutions for treating water at the source, whether in small communities, individual buildings, or industrial settings. This extended exploration delves into the operational mechanisms, key players, and the inherent challenges of decentralized water treatment systems.

Operational Mechanisms and Advantages: Decentralized water treatment systems function independently of a central municipal system, treating water close to where it is used. This proximity to the point of use reduces the need for extensive distribution networks, minimizes water loss, and enhances the efficiency of water use. The benefits of decentralized systems include sustainability and resource conservation, resilience and flexibility, and customization to local needs. (7)

Leading Companies and Innovations: Ecolab, based in the US, is at the forefront of developing decentralized water treatment solutions tailored for commercial and industrial applications. Their advanced systems focus on optimizing water use efficiency and reducing environmental impact. (8) Veolia, headquartered in France, is another key player in this sector. They offer a range of decentralized solutions, from small-scale wastewater treatment plants to modular water treatment units, adaptable to various settings and requirements. (9)

Challenges and Risks: Despite their advantages, decentralized water treatment systems face several challenges: maintaining consistent water quality, integrating with existing infrastructure, navigating regulatory and compliance issues, overcoming financial and technical barriers, and gaining public perception and acceptance. (10)

Future Directions and Potential: The future of decentralized water treatment systems lies in overcoming these challenges through technological innovation, policy development, and public education. Advances in sensor technology, remote monitoring, and smart water management systems can enhance the efficiency and reliability of these systems. Policymakers need to develop supportive frameworks that encourage the adoption of decentralized systems, including financial incentives and clear regulatory guidelines. Additionally, educating the public about the benefits and safety of decentralized water treatment is crucial for widespread acceptance and adoption. (11)

Water Recycling and Reuse Innovations

Water recycling and reuse represent a pivotal shift in sustainable water management, conserving precious water resources and reducing reliance on traditional water supplies. This thorough exploration examines the technologies, leading companies, challenges, and potential solutions in the field of water recycling and reuse.

Technological Advances and Key Innovations: Advancements in water recycling technologies have made it possible to treat and reuse water for various purposes, ranging from agricultural irrigation to industrial processes and even potable uses. These innovations include advanced filtration systems, membrane bioreactors, and treatment processes that remove contaminants to safe levels. (12) Key innovations in this field include greywater treatment systems, advanced membrane technologies, and smart monitoring systems. (13)

Leading Companies in Water Recycling and Reuse: Aquatech, based in the US, specializes in water purification technology, including recycling and reuse. Their systems are designed for efficiency and reliability, catering to industrial and municipal needs. (14) IDE Technologies, headquartered in Israel, is renowned for its innovative water treatment solutions. They focus on developing cost-effective and environmentally friendly water recycling systems, widely used in water-scarce regions. (15)

Challenges in Water Recycling and Reuse: public acceptance of recycled water, economic viability, regulatory hurdles, infrastructure integration, and environmental concerns are significant challenges in this field. (16)

Potential Solutions and Future Directions: public education and awareness campaigns, cost-reduction strategies, policy development and incentives, collaboration with stakeholders, and advancements in technology are crucial for the successful implementation of water recycling and reuse initiatives. (17)

In conclusion, the future of water management is ripe with innovation and potential. From nanotechnology and AI to decentralized systems and water recycling, the advancements in this field hold the promise of a more sustainable and efficient approach to water management. However, these innovations also bring forth new challenges that require careful consideration and collaborative efforts to overcome.

As we move forward, it is imperative that we address these challenges head-on, ensuring that the benefits of these technologies are realized while minimizing their risks. The journey toward a sustainable water future is complex, but with continued innovation, collaboration, and a commitment to responsible management, we can navigate these waters successfully.

Footnotes:

(1) Qu, Xiaolei, et al. "Nanotechnology for a Safe and Sustainable Water Supply: Enabling Integrated Water Treatment and Reuse." Accounts of Chemical Research

(2) Savage, Neil, and Marc Desmulliez. "Water Treatment: Nanofiltration."

(3) NanOasis Technologies, Inc. "NanOasis: Revolutionizing Water Purification."

(4) NanoSun Ltd. "Portable Water Purification Using Nanotechnology."

(5) Keller, Arturo A., et al. "Global Life Cycle Releases of Engineered Nanomaterials." Journal of Nanoparticle Research

(6) Alpaydin, Ethem. "Introduction to Machine Learning." MIT Press, 2020.

(7) United States Environmental Protection Agency. "Decentralized Wastewater Treatment Systems."

(8) Ecolab Inc. "Innovative Water Treatment Solutions."

(9) Veolia Water Technologies. "Decentralized Water Treatment Solutions."

(10) Libralato, Giovanni, et al. "To Centralize or to Decentralize: An Overview of the Most Recent Trends in Wastewater Treatment Management." Journal of

Environmental Management.

(11) United Nations World Water Assessment Programme. "The United Nations World Water Development Report 2021

(12) Asano, Takashi, et al. "Water Reuse: Issues, Technologies, and Applications." McGraw-Hill, 2007.

(13) Ghunmi, Laith A., et al. "Greywater Reuse: Technologies, Applications, and Limitations."

(14) Aquatech International LLC. "Water Purification Technology for Industrial and Municipal Use."

(15) IDE Technologies Ltd. "Water Recycling Solutions."

(16) "Quality of Individual Domestic Greywater Streams and Its Implication for On-Site Treatment and Reuse Possibilities." Environmental Technology.

(17) United States Environmental Protection Agency. "Guidelines for Water Reuse."

CHAPTER 14: THE HEROES AMONG US: EARLY ADOPTERS AND THEIR IMPACT

The transformation in global water safety led by Clear is a testament not just to technological advancement, but more importantly, to the early adopters of this technology. These individuals, managing a diverse range of facilities from hotels to residential buildings, are the vanguards in a pivotal movement. Their commitment to water safety sets a significant precedent, inspiring others to join this noble cause and contributing to the global effort to ensure clean and safe water for everyone.

As we stand today, the world faces a severe water crisis that threatens the health and well-being of millions. Approximately one in three people globally lacks access to safe drinking water. This crisis is exacerbated by factors like climate change, pollution, and inefficient water management, leading to health problems, impeding economic development, and resulting in environmental degradation.

Tackling this crisis is not just a matter of technological solutions but also requires a collaborative approach encompassing NGOs, governments, international bodies, and private entities. Organizations like Water.org, The Water Project, and Charity: Water, spearheaded by visionaries such as Gary White, Matt Damon, and Scott Harrison, are at the forefront, providing innovative, sustainable solutions. These organizations have revolutionized access to safe water, especially in regions like sub-Saharan Africa, through sustainable water projects and comprehensive education on sanitation and hygiene. (1)

Governments and policy makers play a crucial role in managing water resources and implementing effective policies, especially in regions severely affected by water scarcity. These policies regulate water usage, protect water sources from pollution, and promote sustainable water management practices. Investment in infrastructure that ensures safe water delivery and proper wastewater treatment is a critical responsibility of governments worldwide.

The private sector's role in water management is increasingly important. Companies like Xylem and Veolia lead in water technology and services, offering solutions ranging from advanced water treatment systems to intelligent water management tools. These corporations collaborate with governments and communities to implement effective water projects and initiatives, playing a crucial role in addressing global water issues.

Organizations like the United Nations, World Health Organization (WHO), and UNICEF play a crucial role in global water safety and security. They provide research, set international standards, and coordinate efforts among different countries and organizations. Their work in monitoring progress on drinking water, sanitation, and hygiene (WASH) is essential for understanding and addressing global water and sanitation issues. (2)

The Early Adopters: Catalysts of Change: Amid this global effort, the early adopters of Clear technology have emerged as crucial catalysts of change. They are the pioneers in the fight for water safety, setting a shining example for their communities and beyond. Their commitment transcends the immediate benefits of cleaner, safer water; it represents a holistic approach to public health, environmental protection, and sustainable living.

These early adopters are more than just customers; they are partners in a mission to transform how communities around the world access and use water. By installing Clear's advanced water treatment systems, they are taking a proactive stance against waterborne diseases and contamination. Their actions demonstrate a deep understanding of the interconnectedness of health, environment, and technology.

Celebrating Community Heroes: The efforts of these pioneers should be globally recognized. Envisioning international forums and awards that honour these early adopters for their foresight and leadership in water safety is not far-fetched. Such recognition would validate their efforts and inspire others to take similar actions.

Fostering a Global Movement: The early adopters of Clear technology are integral to a larger movement toward sustainable water management and public health. By highlighting their success stories, we can encourage a global shift in how water safety is perceived and prioritized. Their journey can serve as a blueprint for other communities and organizations seeking to make a difference in water safety.

A Unified Effort for a Healthier Future

As we look ahead, the role of these early adopters in shaping the future of global water safety cannot be understated. They are at the forefront of a movement toward healthier, safer communities. Their actions today are laying the groundwork for a future where access to clean and safe water is a universal right.

Footnotes:

(1): Water.org. "Our Impact." Water.org, 2021.

(2): WHO/UNICEF Joint Monitoring Programme. "Progress on drinking water, sanitation and hygiene: 2017 update and SDG baselines." World Health Organization and the United Nations Children's Fund, 2017.

CHAPTER 15: WATER'S IMPACT: THE HIDDEN CURRENTS OF REAL ESTATE VALUE

The relationship between water quality and real estate prices is a crucial yet often overlooked aspect of property value. In this chapter, I delve into the significant correlation between buildings suffering from hot boiling advisories or periodic water contamination and the fluctuation of real estate prices within those same buildings.

The Impact of Water Contamination on Real Estate Prices

Water contamination can lead to hot boiling advisories, which are not uncommon in some states. Florida, in particular, is one of the most affected areas. This phenomenon is often attributed to biological contamination or water contamination resulting from a failing infrastructure. Specifically, the distribution systems in these areas are old and prone to biofouling, which worsens over time. In Florida, frequent floods exacerbate the situation, causing water to infiltrate the distribution systems, leading to pipe bursts and biological contamination. Common contaminants include E. coli and other bacteria. (1)

The correlation between water contamination and real estate prices is evident in the resale price graphs, where all leases experience a downturn during a hot water boiling advisory. Once the contamination is addressed, prices tend to stabilize or even rise. This trend highlights the importance of maintaining high water quality standards in residential and commercial properties.

Case Studies: The Real-Life Impact of Water Contamination

Luxury Condominium Tower in West Palm Beach: A brand-new luxury condominium tower in West Palm Beach faced a significant issue when one of the owners, who had put his condo up for sale for $3.5 million, encountered a roadblock during the due diligence period. The prospective buyer discovered biological contamination in the water and subsequently backed out of the deal. The owner then pursued legal action against the management company and developer, accusing them of negligence in providing contaminated water in a new building, which resulted in a lost sale.

Luxury Condominium Tower in Fort Lauderdale: Another case occurred in a luxury condominium tower in Fort Lauderdale. Clear's team called to assess the situation due to a recurring boiling advisory. Our findings revealed that the building suffered from drops in real estate prices due to ongoing contamination. This situation underscores the importance of regular water quality assessments and timely interventions to prevent long-term impacts on property values.

Suburban Neighbourhood in Michigan: In a suburban neighbourhood in Michigan, residents experienced a decline in property values after a series of water contamination incidents. The local water supply was found to contain high levels of lead, prompting a public health advisory. The news of the contamination spread quickly, leading to a decrease in demand for homes in the area and a subsequent drop in property values. (2) This case highlights the broader implications of water contamination on community well-being and real estate markets.

The Healthy Building Movement: A Global Shift Toward Wellness and Sustainability

In recent years, the real estate industry has witnessed a significant shift toward the healthy building movement. This global trend emphasizes the importance of creating spaces that not only support the physical well-being of occupants but also promote environmental sustainability. Two key certifications that have emerged as benchmarks in this movement are the WELL Certification and the Green Building Certification.

WELL Certification: Pioneering Health and Well-being in Buildings

The WELL Certification is a performance-based system that measures, certifies, and monitors features of the built environment that impact human health and well-being. Developed by the International WELL Building Institute (IWBI), this certification focuses on seven core concepts: air, water, nourishment, light, fitness, comfort, and mind. Buildings that achieve WELL Certification demonstrate a commitment to enhancing the health and wellness of their occupants. (3)

For example, a commercial office building in New York City achieved WELL Certification by incorporating advanced air filtration systems, ergonomic design elements, and biophilic design principles that connect occupants with nature. As a result, the building saw an increase in tenant satisfaction and demand, leading to higher occupancy rates and rental premiums.

Green Building Certification: Embracing Environmental Sustainability

Green Building Certification, such as LEED (Leadership in Energy and Environmental Design), focuses on promoting sustainability in the construction and operation of buildings. These certifications assess various aspects of sustainability, including energy efficiency, water conservation, and the use of sustainable materials.

A residential development in Vancouver, Canada, received LEED Platinum certification for its innovative use of renewable energy sources, water-efficient landscaping, and environmentally friendly construction materials. This certification not only reduced the development's environmental impact but also attracted environmentally conscious buyers, resulting in a rapid sell-out of the units at premium prices. (4)

Impact on Real Estate Reputation and Revenues

The adoption of WELL and Green Building Certifications has a profound impact on the reputation and revenues of real estate properties.

Buildings that achieve these certifications are often perceived as more attractive and prestigious, appealing to a growing segment of health-conscious and environmentally aware consumers and tenants. This increased demand can lead to higher rental rates, faster lease-up times, and enhanced property values.

Furthermore, certified buildings often experience lower operating costs due to energy and water efficiency measures, which can further improve the financial performance of the property. In the competitive real estate market, these certifications serve as a differentiator, setting properties apart and positioning them as leaders in the healthy building movement.

In conclusion, the healthy building movement is reshaping the real estate landscape. As awareness and demand for health and sustainability continue to grow, developers and property owners who embrace these principles are likely to see long-term benefits in terms of reputation, occupancy, and financial returns. By addressing water quality issues proactively and investing in health and wellness features, the real estate industry can contribute to the well-being of occupants and the environment while enhancing property values and market competitiveness.

Rethinking Real Estate: The Clearspaces Miami Revolution

In a bold move to not just promote the vision of healthy buildings but to demonstrate its tangible benefits, I embarked on a transformative project in downtown Miami. I acquired a 16,000 sq ft office complex at 151 SE 1st St., which had remained vacant since its completion five years prior, due to the developer's unsuccessful attempts to lease it for various uses, including offices, medical offices, and gyms. Seizing the opportunity presented by this distressed situation, I acquired the shell at an attractive price per square foot, with the aim of reimagining the office space to meet the evolving needs of post-COVID-era employees.

The office sector across North America has been in a state of decline, with giants like Brookfield and Blackstone offloading office buildings into receivership. The traditional office model, characterized by cramped cubicles and minimal socialization spaces, has become outdated and has led to issues such as noise pollution, decreased efficiency, and an increase in sick days. The pandemic further highlighted the feasibility of remote work, making the return to the traditional office setting even more challenging.

At Clearspaces Miami, I envisioned an ecosystem—a biosphere of health and well-being for office employees. From the moment they enter the lobby, they are greeted by organic natural materials, an indoor forest of banana and fig trees, and elements of wood and stone. On the office floors, I implemented a clear notification system and installed Clear air purifiers that continuously measure six different elements and pollutants in the air, introducing fabric biotics to dramatically reduce harmful bacteria. (5)

To further enhance the well-being of employees, I incorporated a Pilates studio and a virtual reality relaxation room. I also created social spaces with balconies to foster interaction and relaxation. My innovative approach to office design proved to be a success, as I managed to lease 85 percent of the space to leading companies.

The success of Clearspaces Miami serves as living proof that rethinking the office environment and prioritizing health and well-being can lead to better leasing and sales outcomes in today's market. (6) My project demonstrates that healthier, more user-friendly spaces are not only in demand but are also essential for the modern workforce, redefining the future of office spaces in the post-pandemic era.

Health and longevity are not just goals; they are the foundation upon which we build a life of fulfillment and purpose. By prioritizing our well-being, we empower ourselves to achieve our greatest potential and leave a lasting legacy.

Footnotes:

(1): United States Environmental Protection Agency. "Drinking Water Contaminants – Standards and Regulations."

(2): Gómez-Víquez, L., Rios, V., & Ramírez, A. (2020). "Impact of water quality on urban real estate prices: Evidence from Cartagena de Indias, Colombia." Land Use Policy, 99, 104863.

(3): International WELL Building Institute. "WELL Building Standard."

(4): U.S. Green Building Council. "LEED Certification."

(5): Clearspaces. "Innovative Office Solutions for a Healthier Work Environment."

(6): Real Estate Market Trends. "The Rise of Healthy Office Spaces Post-COVID."